The Renewable Energy Handbook for Homeowners

The Complete Step-by-Step Guide to Making (and Selling) your own power from the sun, wind and water

William H. Kemp

AZTEXT PRESS

Tamworth, Ontario
www.aztext.com
cam@aztext.com

Distributed By:
Hushion House Publishing Limited
36 Northline Road, Toronto, Ontario Canada M4B 3E2
Telephone: (416)-285-6100 E-mail: jbeau@hushion.com

National Library of Canada Cataloguing in Publication

Kemp, William H., 1960-
    The renewable energy handbook for homeowners : the complete step-by-step guide to making (and selling) your own power from the sun, wind and water / William H. Kemp.

Includes bibliographical references and index.
ISBN 0-9733233-0-2

    1. Renewable energy sources—Handbooks, manuals, etc. I. Title.

TJ808.3.K44 2003   621.042   C2003-904237-5

Disclaimer:
"The installation and operation of renewable energy systems involves a degree of risk. Ensure that all proper installation, regulatory and common sense safety rules are followed. If you are unsure of what you are doing, STOP! Seek skilled and competent help by discussing these activities with your dealer or electrician.

Electrical systems are subject to the rules of the National Electrical Code™ in the United States, and the Canadian Electrical Code™ in Canada. Wood and pellet stoves are also subject to national and local building code requirements.

In addition, local utility, insurance, zoning and many other issues must be dealt with prior to beginning any installation. A word to the wise, "People don't plan to fail, they just fail to plan." Nowhere is this more apparent than when installing renewable energy systems.

The author and publisher assume no liability for personal injury, property damage, consequential damage or loss, including errors and omissions, from using the information in this book, however caused.

The views expressed in this book are those of the author, and do not necessarily reflect the views of contributors or others who have provided information or material used herein.

Printed and bound in Canada by Marc Veilleux Imprimeur, Boucherville, QC.
Printed on "Enviro", 60 lb, 100% post-consumer recycled paper using vegetable-based inks.

Table of Contents

Table of Contents

Preface

The majority of people don't give much thought to energy. Sure, everyone gripes a bit (maybe a lot) when the cost of electricity or gasoline rises for the tenth time in a year. But do most people really ever think about energy? Where does it come from, what impact does it have on the family budget, the environment or our children's future?

Most of us are so busy trying to stay afloat in our hectic lives that these questions only surface as polite dinner discussions. Every once in a while, someone stops to ponder these issues. Perhaps that's why you have picked up *The Renewable Energy Handbook for Homeowners*.

Within the covers of this handbook, we will provide you with the background information, purchasing data and step-by-step instructions to allow you to:

- Stretch your heating and utility dollar
- Make the electrical meter run backwards by generating your own power
- Provide heat and electricity to that beautiful cottage or home building lot that is just a bit too far from the hydro poles
- Let you worry just a little bit less about acid rain, ozone depletion, airborne contaminants, smog, nuclear waste or world oil supply problems.

Whether you are just curious or an industry expert, if these issues are on your mind, then this handbook is for you.

We will discuss how to stretch your energy dollar, doing much more with less. Step-by-step guidance and easy to understand instruction will help you pave the way to better energy management and renewable energy production, whether or not you cut the electrical supply lines to your home or cottage.

We will guide you to a new relationship with the environment that lets you consume energy without worrying about the effects of dumping today's pollution on tomorrow's children. Unlike fossil fuels or nuclear energy, renewable sources are just that, renewable. No matter how much energy we capture and use, there will always be more for everyone else.

And most important of all, *The Renewable Energy Handbook for Homeowners* will show you how to do it all, now!

Acknowledgements

A book of the magnitude of *The Renewable Energy Handbook for Homeowners* requires the technical expertise of and access to information from many individuals and companies, not to mention the patience of family while writing and editing the text.

It is customary to identify those contributors who have helped with the technical data required to put this book together, and make sure the information is the most up-to-date.

I am indebted to, and would like to thank, *Michael Reed* of Array Technology Inc. for information on tracking solar mounts; *Mike Bergey* of Bergey Wind Power for technical publications and photographs on wind power systems; *Don Bishop* of Benjamin Heating Products for information regarding dual energy heating boilers; *Colin Mitchell* and *Annie Lee* of Dankoff Solar Products for data regarding high efficiency water pumping; *Jody Graham* at Energy Systems and Design for insight on micro-hydro turbine units; *George Peroni* at HydroCap Corporation for information regarding hydrogen gas reclaiming devices for batteries; *Jack Csenge* at Jack Rabbit Energy Systems for photographs and insight into their "run of river" hydro turbine unit; *Patrice Feldman* at Morning Star Corporation for her quick response regarding PV controller design and installation; *Rajindar Rangi* and *Tony Tung* of Natural Resources Canada, Energy Technology Branch, for guidance and providing access to reams of technical bulletins and contacts; *John Supp* of South West Wind Power for the great site photographs and micro-wind technical data; *Govindh Jayaraman* of Topia Energy Inc. for assistance with BioDiesel Fuels; *Pam Carlson* of Xantrex Technology Inc. for the tremendous support in providing technical documentation, photographs and the nice Xantrex coffee mug!; and lastly, *Stefani Kuykendall* of Zomeworks Corp. for information on fixed PV mounting hardware.

I am also indebted to those who have had to work directly with me in the preparation of the text. A big thanks goes out to: *Carol McGregor* for her excellent line drawings and graphics, not to mention having to work under Bill's demanding time lines; *Leann Thompson* for her great help behind the camera. Her work proves that a picture is worth a thousand (of Bill's) words (Leann may be contacted at blackrockphotgraphs@hotmail.com); *Carolyn Woolmer* who worked wonders stopping me from writing too many run-on sentences; To *Karen* and *Jamie Wilson*, *Hillary Houston* and *Raymond Lebfrevre*, the *Adams'*, *Vickie* and *John* at Fox Lake Resort, *Lorraine Kemp* and *Michelle* and *Cam Mather* for letting me invade their privacy to extract the photographs of R.E. systems throughout the book.

And of course thanks to *Lorraine Kemp* and *Michelle Mather* (and *Katie* and *Nicole*) for putting up with *Cam Mather* and me during our single-minded focus on writing and producing this book, using **100% Renewable Energy, of course!** (Cam's company Aztext Electronic Publishing Ltd. is available for your needs; contact cam@aztext.com.)

About the Author

The Author with friends.

William Kemp is Vice President of Engineering for an energy sector protection and controls company. He is actively involved in environmental and electrical safety committees as well as teaching others about renewable energy. Bill and his wife Lorraine have lived in their off-grid hobby/horse farm near Ottawa, Ontario for nine years.
He may be contacted at: whkemp@magma.ca

Introduction

The process of capturing and using renewable energy is really as old as mankind. Sailboats and windmills have captured the wind for centuries, and wood has heated our caves and houses even longer.

Modern renewable energy systems have improved as technology has progressed, becoming more efficient and at the same time more complex. Flip through catalog or web pages for renewable energy equipment and you will find hundreds of specifications and alarming amounts of data presented, offering the latest gadgets to gather energy from the sun. Between the bewildering array of devices and costs, it's no wonder people stay plugged into the coal-fired grid!

In the words of Dr. David Suzuki; " *The Renewable Energy Handbook for Homeowners* provides the tools necessary to guide North Americans in their quest for environmental stewardship and energy self-sufficiency. Energy conservation and renewable energy production are the keys to controlling global climate change and ensuring a sustainable future for all."

Handbook Organization

This handbook is written without regard to the readers' background skills and philosophy. Where does one start talking about the technology and systems required for installing and living with renewable energy? At the risk of writing too much information, the most obvious place to start is the beginning. This may sound obvious, but it is surprising to see how many texts have been written where essential facts and basic information are missing.

To prevent this from happening, the handbook has been organized into the following chapters, starting with the basics and moving to the conclusion, step-by-step.

Chapter 2 answers the question "What is Energy?" and reviews some *basic* mathematics required to understand how it is captured and used in every day life. If you have never thought about amps or volts, or are worried about arithmetic, don't.

This chapter keeps it simple and provides a firm foundation for later discussion on power production and equipment ratings.

Chapter 3 deals with energy conservation; doing more with less. While many consider conservation akin to suffering or doing without, it isn't. Consider two identical homes in the same neighborhood. One home costs $300 per month to heat, light and power; the other $100. How is this possible? Aside from one having a teenager that enjoys hour-long showers, it is possible to provide all of the same energy needs and wants, without breaking the bank.

Chapter 4 takes a close look at heating and cooling with renewable energy. This chapter outlines the types of energy systems available as well as how to select and install them. Because renewable energy systems are variable and require some tending, we will also discuss heating systems using more traditional means.

Chapter 5 is where the fun starts. This chapter introduces you to renewable energy systems and how they produce electricity. The overview discusses both electrical grid-connected and off-grid system configurations. Further chapters look at each individual component of the system and detail their selection and installation.

Chapter 6 discusses how to turn sunlight directly into electricity using photovoltaic cells. The chapter discusses module, panel and array theory, mounting and sun location.

Chapter 7 explains how to make electricity from the wind using modern wind turbines. The primary focus is on wind systems and towers designed specifically for the homeowner, typically under 10 kW peak production.

Chapter 8 is for those of you lucky enough to have a sufficient source of water to produce electricity with micro-hydro generation. This section deals with site evaluation, turbine selection and installation.

Chapter 9 discusses how to save your hard earned energy for a rainy day by banking your electricity in a storage battery system. While those of you operating a grid-interconnected system don't require battery storage, you might wish to review this chapter. Even the best grid-interconnected system doesn't run when the grid goes down!

Chapter 10 is for those times of the year when your renewable energy system is producing more energy than you can use and store. Yes, this happens quite often! The voltage regulation systems monitor and control the charging of your batteries and can even help with hot water heating to boot.

Chapter 11 deals with how to get the juice stored in the batteries into the house wiring. Most people are aware that battery power is different from what comes out of your household wall outlets. The amazing device that performs this conversion is called an inverter. Grid-interconnected systems also use inverters to pump the renewable energy out to the utility system.

Chapter 12 discusses what to do if the sun doesn't shine or the wind doesn't blow for long periods of time. Even grid-interconnected homes become cold and dark, when the local nuclear plant sputters to a stop. Backup energy sources provide insurance in these instances. In this chapter, we review the options available, and what type of equipment is required.

Chapter 13 deals with some of the specifics of Electrical Codes and safety issues. Using renewable energy is just as safe as utility supplied power. The difference is that you will own, manage and maintain your very own power station! This extra responsibility requires a few additional common sense safety steps, which will be reviewed.

Once you have an understanding of safety codes, we can introduce the actual wiring required. Electrical wiring for the *normal* house is standard stuff and well understood by all electricians. For renewable energy equipment, there are some extra rules and common sense issues that must be followed. If you aren't planning on doing the wiring yourself, it might be a good idea to give a copy of this chapter to your electrician.

Chapter 14 will wrap up the discussions in the Handbook and provide a little insight on how to live with renewable energy once you throw the switch and the system is up and running. Living with renewable energy is really quite simple. Perhaps the biggest adjustment is learning how not to gloat the next time your neighbor calls on that dark and stormy night to tell you the power is out again!

Chapter 15 is the conclusion of the book and offers a look into the future of renewable energy for homeowners.

Appendices contain resource information supplementing the chapter data.

A word about pricing. Many of the products described in this handbook are either manufactured in the United States or priced in U.S. dollars. To avoid confusion, all indicated prices are therefore shown in U.S. dollars.

2
What is Energy?

All of the earth's energy comes from the sun. In the case of renewable energy sources and how we harness that solar energy, the link is often very clear: sunlight shining through a window or on a solar heating panel creates warmth, and when it strikes a photovoltaic (or PV) panel the sunlight is converted directly into electricity; the sun's energy causes the winds to blow, which moves the blades of a wind turbine, causing a generator shaft to spin and produce electricity; the sun evaporates water and forms the clouds in the sky from which the water, in the form of raindrops, falls back to earth. The rain falling in the mountains becomes a stream that runs down hill into a micro hydroelectric generator.

While these energy sources are renewable, they are also variable and intermittent. The sun generally goes down at night and may not shine for several cloudy days. If we just wanted to use our energy when it is available, the various systems used to collect and distribute it would be a whole lot simpler. But we humans are just not that easily contented. I know for a fact that most people want their lights to turn on at night, even though the sun stopped shining on the PV panels hours ago. In order to ensure that heat and electricity are available when we need it, a series of cables, fuses, boilers and a

Figure 2-1. Most people want their lights to turn on at night. This little issue complicates the electrical systems of a renewable energy system.

bewildering array of components are required to capture nature's energy and deliver it to us on demand.

The process of capturing and using renewable energy may seem far too complicated and expensive for the average person. While the components themselves are complicated, the theory and techniques required to understand, install and live with renewable energy are not.

Why are we discussing math?

Although it is not absolutely necessary to be an expert in heating, wind and electrical energy, an understanding of the basics will greatly assist you in operating your renewable energy system. This knowledge will also help you make better decisions when it comes to purchasing the most energy efficient appliances or deciphering your previously undecipherable hydro bills.

If you are planning a grid-interconnected system, you may consume as much energy as you like, provided you don't mind paying for it. Whereas for off-grid systems, energy consumption is an important consideration as additional energy comes from more expensive and complex alternative sources or fossil fuel generators.

Trying to save money by reducing energy costs can only be accomplished by understanding your "energy miles per gallon" quotient. It is fairly simple to determine gas fuel economy or efficiency, but it is much more difficult to determine energy efficiency for a house full of appliances and heating equipment.

The mathematical theories described in this *Handbook* are simplified. Equations are used throughout the text, but they are limited to those areas that are most important. While the mathematics might be a bit light for some, for the rest of us, the math and logic used are fairly straightforward; a simple calculator (powered by the sun of course) will make the task that much simpler. For those so inclined, there are plenty of references to sources of data that will make even the most ardent "techie" happy.

The Story of Electrons

If you can remember back to your high school days in science class, you may recall that an atom consists of a number of electrons swirling around a nucleus. When an atom has either an excess or lack of electrons in comparison to its "normal" state, it is negatively or positively

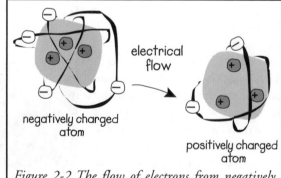

Figure 2-2 The flow of electrons from negatively charged atoms to positively charged ones is known as the flow of electricity.

charged, respectively.

In the same way that two magnets' north and south poles are attracted, two oppositely charged atoms are also attracted. When a negatively charged atom collides with a positively charged one, the excess electrons in the negatively charged atom flow into the positively charged atom. This phenomenon, when it occurs in far larger quantities of atoms, is called the flow of electricity.

The force that causes electricity to flow is commonly known as *voltage* (or V for short). The actual flow of the electrons is referred to as the *current*. So where does this force come from? What makes the electrons flow in the first place? The trigger that brings about the flow of electrons can come from several energy sources. Typical sources are chemical batteries, photovoltaic cells, wind turbines, electric generators and the up-and-coming fuel cell. Each source uses a different means to trigger the flow of electrons. Waterfalls, coal, oil or nuclear energy are commonly used to generate commercial electricity. Fossil or nuclear fuels are used to boil water, which creates the steam that drives a turbine and generator. Falling water drives a turbine and generator directly; the spinning generator shaft induces magnetic fields into the generator windings, forcing electrons to flow.

Let's use the flow of water as a visual aid to understanding the flow of electricity that is otherwise invisible.

Let's presume that the greater the amount of water that moves past you per

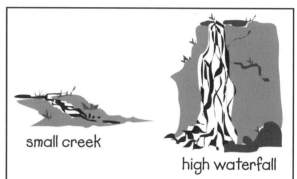

small creek

high waterfall

Figure 2-3 The flow of electricity is very similar to the flow of water. A creek has a lower flow (and pressure) than a waterfall, just as more electrons moving from atom to atom increases electrical current.

second equates to a greater flow, and that a river with a large flow of water has a large current. In electricity, a large number of electrons flowing from one atom to the next are similar to a large flow of water. Therefore, the greater the flow of electrons past a given point per second, the greater the electrical current.

The flow of water is typically measured in gallons per minute or liters per second; electrical current is measured in amperes (or A for short). If the measured current of electrons in a material is said to be 2 A, we know that a certain number of electrons have passed a given point per second. If the current were increased to 4 A, there would be twice the number of electrons flowing through the material.

Gravity is a factor in considering water pressure. A meandering creek has very little water pressure, owing to the small hill or height that the water has to fall.

Whereas a high water fall has a much greater distance to fall and therefore has increased water pressure. For example, if a water-filled balloon were to fall on you from a height of 2 feet you would find this quite refreshing on a hot day. A second balloon falling from a height of 100 feet would exert a much higher pressure and probably knock you out! This increase in water pressure is similar to the electrical pressure or voltage that forces the electrons to flow through a material.

Conductors and Insulators

Electricity flows through conductors in the same manner as water flows through pipes and fittings. When electrons are flowing freely through a material we know that the material is offering little resistance to that flow. These materials are known as conductors. Typical conductors are copper and aluminum, which are the substances used to make electrical wires of varying diameters. Just as a fire hose carries more water than a garden hose, a large electrical wire carries more electrons than thin one. A larger wire can accept a greater flow of current (i.e. handle higher amperage).

Before electricity can be put to use, we must create an electrical circuit. We create such a circuit by connecting a source of electrical voltage to a conductor and causing it flow through a load and back to the source of voltage.

The drawing in Figure 2-4 shows how a simple flashlight works. Electrons stored in the battery (we will cover that one later) are forced by the voltage (pressure) to flow into the conductor from the battery's negative contact (-) through the light bulb and back to the battery's positive contact (+). Current flowing in this manner is called *direct current* (or DC for short). The light will stay lit until the battery dies (runs out of electrons) or until we turn off the switch.

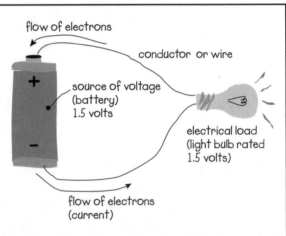

Figure 2-4 An electrical circuit consists of a source of voltage connected to an electrical load through conductors.

Oh yes, a switch would be a good idea. This handy little device allows us to turn off our flashlight to prevent using all of the electrons in the battery. From our description of a circuit, we can assume that by breaking the electrical conductor path, the light will go out.

How do we break the path? By interrupting the flow of electrons using a non-conductive substance. Any substance that does not conduct electricity is known as an insulator.

Typical insulators include rubber, plastics, air, ceramics and glass.

Batteries, Cells and Voltage

You may have noticed that your brand of flashlight has two or even three cells. Placing cells in a stack or in *series*

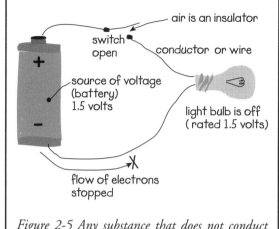

Figure 2-5 Any substance that does not conduct electricity is an insulator.

causes the voltage to increase. For example, the flashlight shown in Figure 2-5 contains a cell rated at 1.5 V. Placing two cells in series, as shown in Figure 2-6, increases the voltage to 3 V (1.5 V + 1.5 V = 3 V). The light bulb in Figure 2-6 is glowing very brightly and will quickly burn out because it is rated for 1.5 V. In this example, a higher voltage (pressure) is causing more current to flow through the circuit than the bulb is able to withstand. Likewise, if the cell voltage were lower than the rating of the bulb, insufficient current would flow and the bulb would be dim. This is what happens when your flashlight batteries are nearly dead and the light is

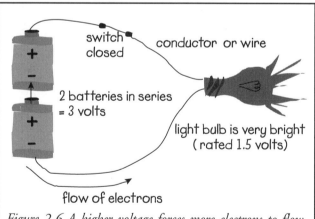

Figure 2-6 A higher voltage forces more electrons to flow (higher current) than the bulb is rated to tolerate. Always ensure that the source and load voltage are rated equally!

becoming dim: the batteries are running out of electrons.

Next time you are waiting in the grocery store line up, put down that copy of the National Enquirer and take a look at the battery display. The selection will include AA, C and D sizes of cell, which all have a rating of 1.5 V. Can you guess the difference between them?

A larger cell holds more electrons than a smaller one. With more electrons, a larger battery cell can power an electrical load longer than a smaller one. The example circuit shown in Figure 2-8 shows a set of jumper wires connecting the cell terminals in *parallel*. This parallel arrangement creates a battery bank of 4 AA style cells, which has the same number of electrons as and capacity of the C size cell. Any grouping of cells, whether connected in series, parallel or both is called a battery bank or battery.

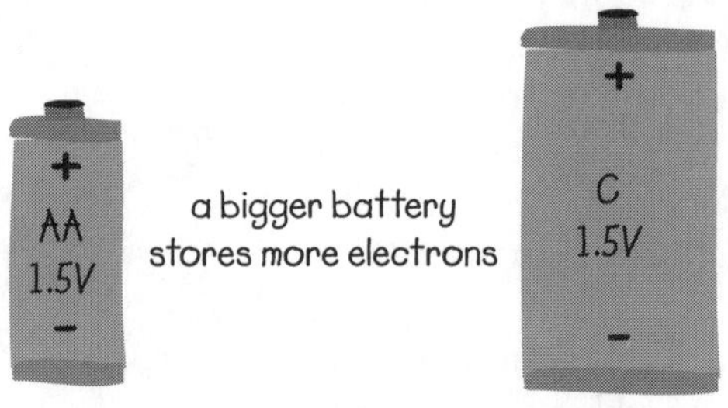

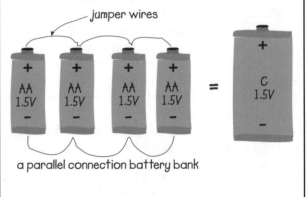

Figure 2-7 Cells of different size contain more electron storage. Just as a 2 liter jug of water contains twice as much water as a 1 liter bottle, a larger battery stores more electrons and works for longer periods.

Obviously, a house requires far more electricity than a simple flashlight does. Off-grid homes generate electricity from renewable sources (more on that later) and store it in battery banks such as the one shown in Figure 2-9.

Figure 2-8 Wiring cells in parallel increases the capacity of the battery bank.

Each battery cell is numbered 1 through 12 and has a nominal voltage of 2 V. If you look carefully, you can see that each cell has a (+) and (-) terminal. Each terminal is wired in *series* to the next battery in the same manner as outlined in Figure 2-6. Therefore a series string of 12 cells rated at 2 V each creates a battery bank rated at 24 V DC (or Vdc for short).

You will also note that there are two such banks of batteries. The bank on the left is identical to the bank on the right. By wiring the two banks in *parallel* we create a total battery capacity that is twice as large as a single bank. This is exactly the same as wiring the 4 AA cells in parallel making the equivalent of the large C cell. Obviously the more electricity we use, the greater the battery size and cost.

Figure 2-9 Off-grid homes require large amounts of electricity, which is stored in a battery bank such as this one.

Grid-Connected Houses Don't Run on Batteries

You are probably aware that your house works on 120/240 V and not 12 V, 24 V or 48 V from a battery. Early off-grid houses, small cottages, boats and many recreational vehicles can and do use 12 Vdc systems. But, don't consider using low-voltage DC for anything but the smallest of systems. The limited selection of appliances and difficulty in wiring a low-voltage, full-time home is not viable.

The modern grid-connected home is supplied with electricity by an electrical utility in the form of 120/ 240 V *alternating current* (or Vac for short). In a DC circuit, as defined earlier, current flows from the negative terminal of the battery through the load and back to the positive terminal of the battery. Since the current is always flowing one way, in a direct route, this is called direct current.

In an AC circuit, the current flow starts at a first terminal and flows through the load to the second terminal, similar to a battery circuit. However, a fraction of a second later, the current stops flowing, and then reverses direction flowing from the second ter-

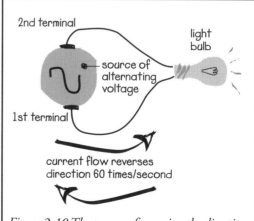

Figure 2-10 The process of reversing the direction of current flow is known as alternating current.

minal, through the load and back to the first terminal as shown in Figure 2-10.

Alternating Current in the Home

Generating electricity in the modern, grid-connected world is accomplished using various mechanical turbines turning an electrical generator. In the early years of the electrical system, there was considerable debate as to whether the generator output should be transmitted as AC or DC. For a time, both AC and DC were generated and transmitted throughout a city. However, as time passed, it became clear for safety, transmission and other practical reasons that AC was more desirable. As the old saying goes, "the rest is history". All modern houses and resulting electrical appliances are standardized in North America to operate on either 120 Vac or 240 Vac. Because of this decision, it is advisable for off-grid homes to convert the low-voltage electrical energy stored in the battery bank to 120/240 Vac. Likewise, for renewable energy systems that are grid inter-connected, it is necessary to convert the DC voltage to 120 Vac or 240 Vac for storage on the grid. The device used for this electrical conversion is the inverter, which we will discuss in more detail in Chapter 11.

Power, Energy and Conservation

Conserving energy and doing more with less is not only good for the planet but helps to keep the size and cost of your renewable energy power station within reasonable limits. We discussed earlier how a bigger battery could run a light bulb longer than a smaller battery. This is fairly obvious. What might not be so obvious is that if we replace the "ordinary" light bulb with a more efficient one, we might not need the larger battery in the first place.

Let's review how any typical electrical circuit operates: A source of electrons under pressure (i.e. voltage) flows through a conductor to an electrical load and back to the source. For household circuits, the voltage (pressure) is usually fixed at

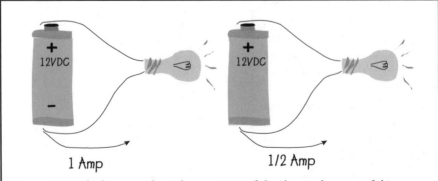

Figure 2-11 The lamp on the right uses 50% of the electrical energy of the one on the left and both have equal brightness. The lamp on the right is said to be twice as efficient.

120 Vac or 240 Vac. For battery supplied circuits, the voltage is usually fixed at either 12 Vdc, 24 Vdc or 48 Vdc, depending on the size of the load and the amount of electron flow (i.e. current) that is required to make the load operate. Let's say that an ordinary light bulb requires 12 V of pressure and 1 A of current flow to make it light. Now suppose that we can find a light bulb that uses 12 V and only 0.5 A of current flow. Assuming that both lights are the same brightness, we can instinctively tell that the second light is twice as efficient as the first. Stated another way, we would need only half the battery bank size (at lower cost) to run the second lamp for the same period of time, or the same size of battery bank would run the second lamp for twice as long.

The relationship between the pressure (voltage) required to push the electrons to flow in a circuit and the number of electrons actually flowing to make the load operate (current or amps) is the power (commonly expressed as *watts* or W for short) consumed by the load. Using the above example, let's compare the power of the two circuits:

Ordinary Lamp:
   12 V x 1 A = 12 W
More Efficient Lamp:
   12 V x 0.5 A = 6 W

As our instincts told us, the second lamp uses half the number of electrons to operate. The power of a circuit is simply the voltage multiplied by the current in amps, giving us the instantaneous flow of electrons in the circuit measured in watts.

Remember that we made the comment that the more efficient bulb operates twice as long on a battery compared to the ordinary one. Where does the time enter into this? If a battery has a known number of electrons stored in it, and we use them up at a given rate, the battery will become empty over time. The use of electrons (power) over a period of time is known as *energy*.

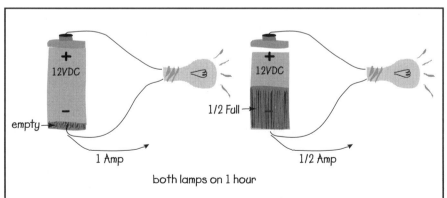

Figure 2-12 The more efficient lamp on the right uses power at a lower rate. Its energy consumption is 50 % lower than the lamp on the left.

Energy is *power* multiplied by the *time* the load is turned on:

Ordinary Lamp:

   12 V x 1 A x 1 hour = 12 Watt hours   or;
   12 W x 1 hour = 12 Watt hours

More Efficient Lamp:

   12 V x 0.5 A x 1 hour = 6 Watt hours   or;
   6 W x 1 hour = 6 Watt hours

Your hydro bill charges you for energy, not power. You run around turning off unused lights to cut down on the amount of time the lights are left running. In an off-grid house, we want to drain the batteries as slowly as possible, using energy with the most efficiency. Turning off unused electrical loads and running appliances that, like our light bulb example, are more energy efficient.

If we think about it long enough, we can also use these calculations to figure out how much energy is stored in a battery bank. For example, if a 12 V battery bank can run a 10 A load for 30 hours, how much energy is stored in the battery?

Battery Bank Energy (in Watt Hours):

   12 V x 10 A x 30 hours = 3,600 Watt hours or 3.6 kilo-Watt hours of energy

An interesting thing about batteries is that their voltage tends to be a bit "elastic". The voltage in a battery tends to dip and rise as a function of its state of charge. For example, our nominal 12 V battery bank only registers 12 V when the batteries are nearly dead. When they are under full charge, the voltage may reach nearly 16 V. Because of this swing in the voltage, many batteries are not sized in *Watt hours* of energy, but *Amp hours*.

In fact, the math is similar; just drop the voltage from the energy calculation:

Battery Bank Energy (in Amp Hours):

   10 A x 30 hours = 300 Amp hours of energy

To convert Amp hours of energy to Watt-hours, simply multiply Amp hours by the nominal battery voltage. Likewise, to convert Watt-hours of battery bank capacity to Amp hours, divide Watt-hours by the nominal battery voltage.

That pretty well covers everything you need to know about electrical energy. With a bit of practice (and you'll get plenty of that in the chapters to follow) you will know this stuff well enough to brag at your next office party.

Heating and Cooling Energy
Heating

If you happen to be sitting beside a nice warm wood stove as you read this *Handbook*, you can feel the heat from the fire radiating out towards you. This warmth and the effects of reading the section above on electrons may have caused you to doze off to sleep. If not, you might consider that heat is not the same thing as electricity, for if it were, it could not reach you on account of air being an insulator.

In fact, the idea that heat is a form of energy baffled many earlier physicists and took a long time to understand. It took an even longer period to transfer the theory of heat into industry. Consider that houses built around the turn of the century gave no thought to conserving heat by means of insulating or any of a number of techniques which seem quite obvious to us now.

So what exactly is heat? Early researchers thought that heat was a substance, something you could put in a bottle: a fluid they called *caloric*. This theory persisted until the middle of the 19th century and was not nearly as silly as you might think. Think about it….heating a liquid causes it to expand, as if something is added to it; when wood is burned, a small pile of ashes is left behind, as if something escaped or evaporated; a jar of boiling hot water placed next to a jar of cool water causes the cool water to warm up. Perhaps *caloric* flowed from the warmer jar to the colder?

Although we now understand that the notion of *caloric* is incorrect, the concept of heat flowing as if it were a fluid is not that far from the truth, as the above examples illustrate. As scientists continued scratching their collective heads, the current laws of heat and thermodynamics slowly replaced the theory of *caloric*.

It is now understood that heat is a form of energy caused by the motion of molecules, or groups of atoms. All matter is made up of molecules, which (like teenagers) are always in motion. As a substance is warmed, its molecules move faster; as it is cooled, its molecules move less. The temperature of a substance is directly related to the motion of these tiny molecules.

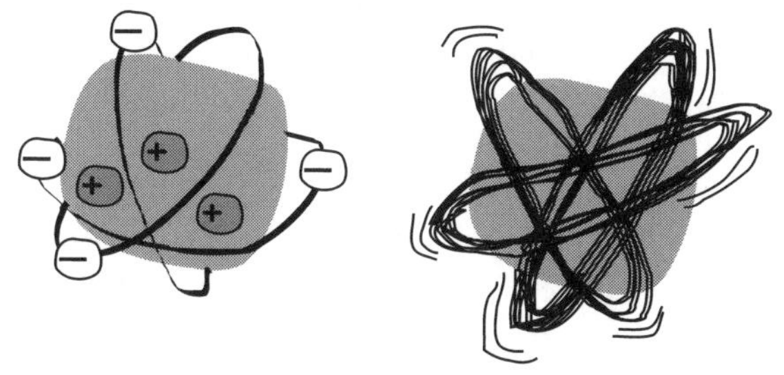

Figure 2-13 The atom on the left is cooler than the one on the right because it is moving more slowly.

One of the primary practical considerations relating to heat is that it does not like to stay still (also like teenagers). Consider any source of heat and you will notice that it always wants to move from the hotter object to a colder one. A hot pan placed in cold water causes the pan to cool, while warming the water. In the same manner, your expensive winter heated air wants to get outside as quickly as possible to help melt the snow. Although melting the snow might be a useful task, the cost in terms of dollars and energy is just a bit beyond the reach of reality.

When we wish to stop the flow of electricity in a circuit, we use a non-conductive device called an insulator. Heat energy can also be slowed down in its relentless path into or out of our homes by the use of an insulator for heat (insulation), which is discussed in Chapter 3.

Because heat is energy, it is also possible to quantify this energy the same way that electricity is quantified in kilowatt-hour units. As a matter of fact, if we were to heat, cool and operate our homes completely on electricity, the energy usage charge would indeed be in kWh as all of the energy used would be sourced from the electrical utility.

Unfortunately, there are many sources of heating/cooling energy to choose from, including natural gas, electricity, propane, oil, wood and wood by-products and even coal. Each of these sources of heat energy is delivered to you in a bewildering array of units, making cross comparison-shopping very difficult. To make matters worse, the efficiency of the heating or cooling equipment using these various sources vary greatly.

To level the playing field, we will use the English system "British thermal unit", or BTU, as our standard unit for comparing heating and cooling energy. For readers more accustomed to the metric system, the calorie or joule measure will provide the same basis for relative comparison. Just make sure that you don't mix both systems at the same time, or you may be wondering why you have to chip ice out of your toilets next January.

One BTU is the quantity of heat required to raise, by one Fahrenheit degree, the temperature of one pound of water. Using these units allow you to quickly compare two heating sources which have differing base units. For example, suppose you are trying to compare the cost of heating a house with oil or propane. Which is more economical?

Assume the current cost to heat your home is $500.00 per year and that you require 250 gallons of oil at $2.00 per gallon. Propane costs $1.75 per gallon. You assume that propane costs 25 cents less per gallon and that it will be cheaper to run. Sorry, it doesn't work like that. The first step is to find out how much heat energy in BTUs is stored in each energy source. Refer to the cross-reference chart in Appendix 1, and look up information on both energy sources. You will see that oil contains 142,000 BTUs per gallon and propane contains 91,500 BTUs per gallon. Therefore:

250 gallons of oil x 142,000 BTU/gallon = 35,500,000 BTU per heating season
is required to heat your house.

And, if we were to assume we need 250 gallons of propane:
250 gallons of propane x 91,500 BTU/gallon = 22,875,000 BTU

But your house requires 35.5 million BTU, so we have a shortfall of:
35,500,000 required – 22,875,000 from propane = 12,635,000 BTU shortfall

Now let's take a look at the costs. The current heating charge using oil is $500.00, and our assumption that 250 gallons of propane would be sufficient produced our first estimate of:

250 gallons of propane x $1.75 per gallon = $437.50 or a savings of $62.50/year

In order to make up the shortfall of 12 million BTU, we have to purchase more propane:

12,635,000 BTU shortfall ÷ 91,500 BTU/gallon propane = 138 more gallons
138 more gallons x $1.75 / gallon of propane = $241.50 additional cost

As you can see, our cost for using propane over heating oil went from an estimated saving of $62.50 to a cost of $679.00 per year, for an increase of $179.00 per year! Of course these costs are not real, but the example shows that the amount of heat energy "stored" in a fuel is known and can be reviewed for comparison, when all fuels use the same base units of BTU or joules. This also applies to renewable sources such as firewood, wood pellets and even solar heating systems.

You must also consider efficiency of the heating appliance. Suppose you find two nearly identical heating devices with output ratings of 120,000 BTUs per hour (the fuel type is not important for this calculation). One has an efficiency of 80% and the other 65%. It stands to reason that the 80% efficiency rated unit is the better buy. Over the lifetime of a product, this difference in efficiency can add up to tens of thousands of dollars.

Cooling

An important consideration for homeowners is how to "make" cool air in the dog days of summer. No matter how big your air conditioner unit is, you do not make cold air. A better way to understand a mechanical cooling system is as a *heat pump*. Modern air conditioning units are complex devices that move heat from one location to another. A window air conditioner pumps heat from your warm indoor room and moves it outside. This may seem a little difficult to understand until you realize that the part of the A/C unit outside is quite hot, due to the indoor heat moved there by the internal mechanism. As the outside component (the condenser) is hotter than the outdoor air, the heat (guess what?) wants to move from the hot condenser to the relatively cooler outdoor air. This phenomenon of air conditioning heat

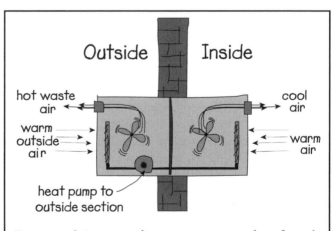

Figure 2-14 An air conditioning unit pumps heat from the room to be cooled and sends it outdoors through a complex compressor system.

transfer explains why the unit has to work harder in hotter weather. As the outside air temperature approximates that of the condenser, heat transfer slows, making the unit run for longer periods. This is also the reason why you want to place the condenser facing north or out of direct sunlight.

This cooling process is not efficient in even the best-designed air conditioners or more complex heat pump systems. Regardless of the type of A/C system installed, the energy used to cool a given area is measured by either the BTUs of heat moved away or by the electrical energy consumed by the unit. Commercial A/C units are usually rated both ways, with some units still carrying a "cooling tonnage" rating. This archaic rating compares the desired A/C units cooling capacity to a ton of ice. Just in case you were wondering, this is approximately equal to 12,000 BTU of cooling capacity.

With all of the heating and cooling energy flowing here and there it seems that just maybe the *caloric* theory really should have been given more consideration after all. At least we could measure the stuff in gallons or liters and not worry about BTUs and cooling tons.

Wind and Streams Have Energy Too

No discussion on renewable energy would be complete without reviewing the energy in the wind and streams. Summer breezes, fresh spring blows and hurricanes are all examples of wind, yet each evokes a very different vision in your minds eye. A gentle breeze has only enough energy to rustle a few leaves, while a hurricane can easily remove the roof off of a house. Same air, same place, what is the difference in strength and why does this happen?

First, lets understand what we mean by wind. Everyone knows that the atmosphere surrounds the earth and is comprised of a colorless, odorless gas, unless you are down wind of a coal fired power plant. The atmosphere that we are concerned with is approximately five miles thick (eight km). In this band, most of our weather patterns occur, caused by the uneven heating effects of the sun. Weather patterns create high and low pressure areas, which cause air to rush between these zones. The greater the difference in the pressure and temperature between these areas, the greater the speed of the moving air, which we call wind.

When we refer to wind as an energy source, we imply we have some means of trapping this energy. Over many hundreds of years, man has found numerous creative means of capturing this energy to do useful work. A large sailboat is an excellent example of putting wind energy to work.

It is obvious that a given size of sail will move the boat faster in high-speed winds than in low-speed winds. Expressed differently, moving the mass of the boat at higher velocity requires higher energy. This seemingly simple fact often eludes many people when it comes to the installation of wind turbine systems. Just as you wouldn't install a hydro dam in a small creek, it is equally inadvisable to place a wind turbine in areas of poor winds. The expression "poor winds" therefore requires some sort of quantification.

In physics, we refer to air as a fluid, just as water is a fluid. Both water and air have mass (or weight if you prefer). When these masses are moving, whether it is water falling under the force of gravity or wind blowing, the fluids contain what is know as kinetic energy. Kinetic energy simply means "the energy of motion". When a surfer rides a wave of water, the water contains sufficient energy to lift and move the surfer along. The kinetic energy of a large river rushing through a hydro-turbine pushes the turbine blades out of the way as the water moves past.

Wind energy works in much the same way; only it is lighter and usually invisible: in this case, the kinetic energy of wind pushes the blades of a wind turbine out of the way.

The amount of kinetic energy then is obviously related to its motion or speed. Equally important is the density of the fluid. Density relates to the mass per unit of volume: a swimming pool

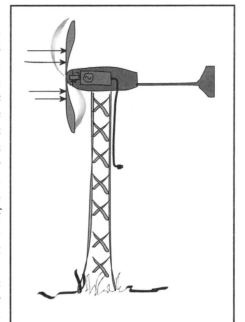

Figure 2-15 The wind blowing through a wind turbine uses its kinetic energy to push the turbine blades out of the way. Some of the kinetic energy is then transformed into mechanical energy, which in turn operates the electrical generator inside the housing.

full of water has much more mass than a swimming pool full of air.

Now lets consider these facts for a moment: if the speed and density is what matters to kinetic energy, then it stands to reason that you need to have the fastest moving fluid, with the highest density to extract the greatest amount of energy. Further, if we could capture more of the moving fluid at one time, we would increase the available kinetic energy further. In practice this is exactly what engineers attempt to do when they install a power station:

- Water has a much higher density than air; therefore water should be the "fluid" of choice. In fact, most of the viable major hydroelectric sites have already been developed in North America. A hydro and wind turbine of similar electrical output requires the wind turbine to be much larger, owing to the lower density of the wind.
- Fluid speed increases power. Major wind farms are located in areas where known wind speed is very high for long periods of time. The Tehachapi Pass in California and Pincher Creek in Alberta are examples of such areas.

- Massive water turbines at Niagara Falls and many wind turbines with over 190 foot (70 m) diameter blades capture ever larger amounts of "fluid" at one time.

So, what units of energy should we use to quantify wind and water energy? If we were concerned with installing multi-million dollar wind farms or another Hoover Dam, energy would be calculated in complex units such as power density in watts per square meter of swept turbine rotor area. Fortunately, for the smaller home-based systems, experience and data tables have provided us with the data necessary to reduce some of this headache.

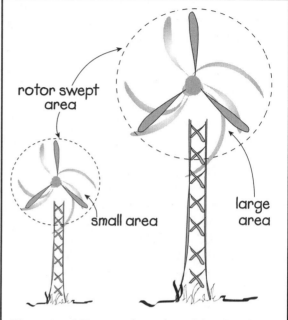

Figure 2-16 Fast wind speeds and large turbine blades increase the energy that can be collected from the wind.

The kinetic energy in the wind is related to its density and speed. The density of wind plays a minor role in the overall calculation unless your site is located at over 3000 feet (1000 m) above sea level. Wind speed, on the other hand, has an exponential relationship to power, meaning that a doubling of wind speed increases available power by eight times! For the purposes of residential-sized wind turbines, we will simply use wind speed coupled with the manufacturers specifications to provide energy figures in watts of electricity. The quantity of wind captured depends directly on the area the wind turbine blades "sweep", and will be discussed in Chapter 7, which covers wind turbine equipment.

Hydro turbines are very similar to wind turbines, with the "fluid" changing from wind to water. There are two variables considered in water energy calculations: *head* and *flow*.

Water *head* is a fancy name for height. The further the water has to fall or run through a vertical drop in a pipe, the larger the head. Consider that a 28-inch long vertical pipe of water exerts a pressure of 1 pound per square inch (or psi for short). A similar pipe 233 feet long exerts a pressure of 100 psi. This additional pressure allows the water to push harder on the turbine, generating more power.

Water *flow* is just that, the flow of water past a given point, measured, for example, in gallons per minute. A higher flow works the same way as higher head, exerting more force on the turbine, with increasing flow. Standing in a small creek offers little obstacle to walking. Try the same thing in a fast moving river; the results are a nice swim as the river overpowers you.

As is the case with wind turbines, using data tables and manufacturers' equipment data sheets allows us to convert *head* and *flow* directly into electrical power in watts, greatly simplifying energy calculations .

Sun-spiration

"Lorraine and I were watching a movie one winters' night a few years ago. The house and yard lights were burning, offering a path to the warmth of our home, at least until the automatic timers closed the lights of access around midnight.

Lorraine heard them first. A few Electrical Utility trucks near the end of our lane way accompanied by snowmobiles. This was a bit odd, but not entirely uncommon as we have a few neighbors further down the road serviced by hydro. The trucks idled patiently while we continued with the movie. Surely if there was a problem, they would come to the door?

Bed beckoned and still the trucks sat there. Oh well, there wasn't much we could do.

The next morning we had a distraught call from one of our neighbors, complaining that we had caused them to freeze in the dark! How so?, I inquired. Apparently the hydro boys were following a broken wire. The houses before us had no power and suddenly ours appeared as an illuminated oasis in the night. Hearing no generator running and fearing their electrical grid maps were incorrect, new ones were summoned from the office about 2 hours away.

None of the repair people believed you could run a house as well as we do, on a renewable energy system."

Bill Kemp

3
Energy Conservation

When the price of gasoline jumps for the tenth time in a month, almost everyone tends to cut back on their driving. Some people might even consider purchasing a vehicle with better gas mileage than a Hummer.

One of the most interesting statistics to be floated about is from the U.S. Department of Energy. They state that for every dollar you spend on conservation, you will save three to five dollars in generation equipment. This applies both to the big nuclear plant built by the mega-utility and your own personal renewable energy power station at home. When energy prices are low, there is little incentive to conserve, notwithstanding environmental concerns. This is especially true when you don't have to pay in advance to help build the utilities' power plants!

This logic doesn't apply when you build your own power station: for every dollar in electricity you save, you can reduce the cost of the installation by up to five dollars. That's a pretty darn good return on investment by anybody's standards.

I am well acquainted with explaining this logic: on Sunday afternoons, tourists often stop by to have a chat about my "big fan on the pole over there". Enjoying a cup of solar power brewed coffee, they mull over why they don't consider using renewable energy too.

At some point in our conversation, I am asked the usual question: "Our hydro bill is about a hundred bucks a month. How can we get it down to zero like you?" It's a good question and the answer is always the same: it doesn't matter whether you want to save a few dollars on your electrical bill or cut the lines to the nuclear plant completely, the path starts in the same place...energy conservation.

Why is energy conservation such a big deal? We just want to make energy, right? Wrong. Dead wrong. Trying to live with renewable energy is destined for failure without first grasping the need for energy conservation. Put very simply, the

Figure 3-1 An energy conserving life style does not mean a Spartan life style. Choosing the most energy efficient appliances that "do more with less" is how to make it happen.

more energy you need to operate your house, the larger the size of power station you must own. It can become very expensive.

Those connected to "Big Coal" don't get off any easier. If home heating and lighting costs keep rising as they have in the last few years, we will all need pockets as deep as Bill Gates.

Doesn't energy conservation mean giving up things; not living the middle class dream North Americans have come to expect; or living a Spartan life of near poverty? Well, yes and no. Yes, energy conservation does mean giving up things, but they are mostly wasteful, inefficient things. What it does not mean is compromising the quality of your lifestyle. And none of this talk even mentions the extra dollars you will be left with after you embrace the lifestyle of an energy conserver.

Figure 3-2 This off grid home is typical of many rural properties, with the exception of the wind turbine, photovoltaic panel and lack of hydro bill.

So, by now you are probably thinking you must reduce your electrical loads before you invest in renewable energy systems. True, but no one wants to give up their TV, stereo, DVD player, computer, coffee machines, hairdryers and the like. And the house has to be warm in winter and cool in summer. How can we reduce our heating and electrical loads to embrace energy conservation without giving anything up?

The answer is efficiency. As the corporate downsizing mantra of late goes: "Do more with less."

Let's start our quest for home energy efficiency with an understanding of the house itself. First, we'll look at some of the most obvious design features that can be incorporated into the building, whether it is on- or off-grid. Then we'll take a look around inside. Lighting, heating, electronics, major appliances all consume energy and must be considered in an energy efficient lifestyle.

New Home Design Considerations

The largest energy usage in North American homes is by far heating and cooling. As we learned in Chapter 2, heat has a nasty habit of wanting to escape outside in winter. In summer, the opposite is true. Heat from the hot summer sun just can't wait to come inside and enjoy the air conditioning with you.

Heat Loss and Insulation

Heat loss and gain in your home is linked to the level and quality of insulation in the ceilings, walls and basement. Not so obvious are the losses associated with windows, doors and sealing in the various joints and holes through the structural cavity. The key to designing an energy efficient house is to ensure that it is well constructed and airtight and contains adequate levels of insulation.

The national and local building codes in your area set the *minimum* stand-

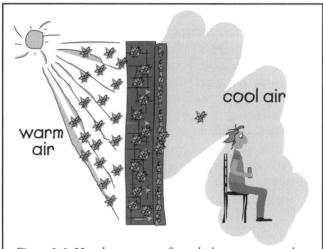

Figure 3-3 Heat loves to move from the hot summer outdoors to enjoy the air conditioning inside with you. Some materials such as brick and stone allow heat to travel fairly easily. Proper insulation slows the flow of heat down.

ards that your building must meet, but it is fairly easy to incorporate upgrades into your design that will pay for themselves many times over. With the world's political climate and uncertainty over fossil fuel supplies in the coming years, these upgrades will provide a safety net against fluctuating energy prices.

Insulation in the walls and ceilings of our homes slows the transfer of heat into or out of our house (in much the same manner that electrical energy cannot flow in a non-conductive material). The higher the quality and thickness of the insulation, the harder time the heat has getting through. Many people believe that because hot air rises, most of the heat loss in the house will be up through the ceiling. No so. Heat moves wherever it can, shifting from warm areas to colder ones, whether that is upwards, downwards or sideways. Keep this in mind as we visit all areas of the home during our insulation spree.

Table 3-1 lists typical recommended insulation values for a very well insulated home. The "R" value (English system) and "RSI" value (metric system) indicate quality levels of insulation. The higher the value, the better the insulation level.

Table 3-1: Minimum Recommended Insulation Values

Insulation Quality (over unheated space)	Walls	Basement Wall	Roof	Floor
R Value	23	13	40	30
RSI Value	4.1	2.2	7.1	5

Moisture Barriers

A vapor barrier consisting of 6-mil thick (0.006 inch) polyethylene plastic can be attached to the wall structure on the warm side of the insulation. The vapor barrier completely surrounds the inside of the house and must be well sealed at the joints and overlapped edges. During construction, care must be taken to ensure this barrier continues from one floor to the next.

The function of the vapor barrier is to seal the house in a plastic bag, controlling air intake and leakage. It also stops warm moist air from penetrating into the insulation system. If warm moist air were to contact cooler air inside the insulation, it would condense, causing mold and rot problems in the wall. Additionally, a vapor barrier ensures the air inside the insulation remains still. These points are absolutely essential in making the insulation system function properly.

Uncontrolled moisture can cause wood rot, peel paint, damage plaster and ruin carpets. Moisture can directly influence the formation of molds, which are allergens to many people. Moisture control is not to be taken lightly. Left uncontrolled, it can cause damage to building components.

Some insulation materials such as urethane foam spray or sheet styrofoam are fabricated with millions of trapped air or nitrogen bubbles in the plastic material.

Although more expensive than traditional fiberglass, inch-for-inch they provide higher insulation ratings and also form their own vapor barrier.

Wind Barriers

A well taped and jointed Tyvek® style wind and water barrier on the outside of the house, just under the siding material, helps to keep winter winds from whistling right through the insulation. Controlling wind pressure infiltration into the building structure is an area most people overlook. Even if the summer breeze isn't blowing through your hair while the windows are closed, wind leakage is important.

Insulation works by keeping dry air very still. Even the slightest movement will greatly reduce the insulation value of the system. Think about how much force you exert wrestling with an umbrella in even a light wind. Imagine now the same affect wind has on the entire surface area of a house. Quite frankly it's amazing more homes don't blow over right now! The force of the wind on the house structure is trying to blow through your insulation. This upsets our still air requirement and lowers insulation values.

Provided all of the above insulation techniques are followed carefully, you will have a very efficient home insulation system.

Examples of the above techniques are shown in the cross-section or side view of the bungalow house in Figure 3-4. This design complements your local building code by increasing insulation in areas that you may not have considered.

Basement Floor

This floor can comprise a 3 inch thick sheet of extruded foam board insulation directly on top of packed crushed stone. Over this can be applied an 8 mil layer of polyethylene moisture barrier, with all joints taped or overlapping seams caulked with acoustical sealant. Such a membrane prevents ground-based moisture from entering through the floor area, reducing mold potential. It will also stop ground gases including radon from entering. In areas where foundation footings are not subject to frost heaving, a footing insulation board can be added between the basement wall and the concrete slab floor.

Basement Walls

The basement walls can be "finished" using standard framing techniques, with the exception that the 2 x 4 inch stud wall is set away from the concrete basement wall by 4 inches. This increases the wall cavity space for additional insulation without utilizing additional framing material. Instead of traditional batt insulation, blown in fiberglass or cellulose is suggested. These materials ensure complete coverage and packing density, especially in the hard to insulate areas around plumbing lines, electrical boxes and wiring. Figure 3-5 shows the use of a tarpaper style moisture barrier glued to the concrete wall from just below grade, which is then folded up and glued to the bottom of the interior vapor barrier.

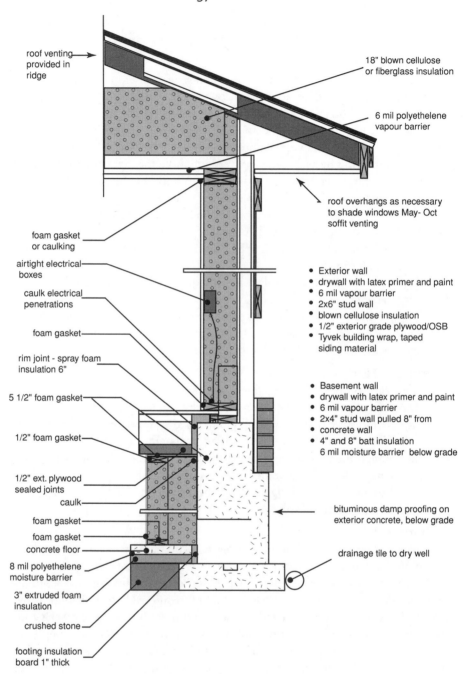

roof venting provided in ridge

18" blown cellulose or fiberglass insulation

6 mil polyethelene vapour barrier

roof overhangs as necessary to shade windows May- Oct soffit venting

foam gasket or caulking

airtight electrical boxes

caulk electrical penetrations

foam gasket

rim joint - spray foam insulation 6"

5 1/2" foam gasket

1/2" foam gasket

1/2" ext. plywood sealed joints

caulk

foam gasket

foam gasket

concrete floor

8 mil polyethelene moisture barrier

3" extruded foam insulation

crushed stone

footing insulation board 1" thick

- Exterior wall
- drywall with latex primer and paint
- 6 mil vapour barrier
- 2x6" stud wall
- blown cellulose insulation
- 1/2" exterior grade plywood/OSB
- Tyvek building wrap, taped siding material

- Basement wall
- drywall with latex primer and paint
- 6 mil vapour barrier
- 2x4" stud wall pulled 8" from concrete wall
- 4" and 8" batt insulation 6 mil moisture barrier below grade

bituminous damp proofing on exterior concrete, below grade

drainage tile to dry well

Figure 3-4 This cross-section of a high-efficiency house illustrates additional insulation and sealing techniques that will greatly reduce your heating and cooling energy usage.

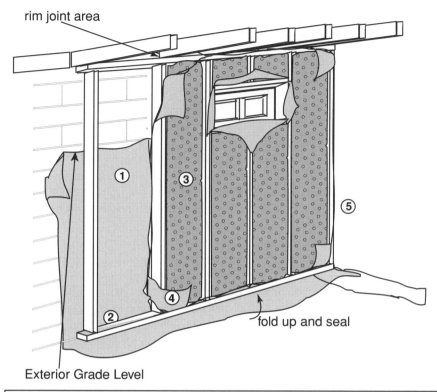

rim joint area

① ③ ⑤ ④ ②

fold up and seal

Exterior Grade Level

Interior insulation involves: 1) a moisture barrier; 2) new frame wall; 3) insulation; 4) air and vapor barrier; 5) finishing

Figure 3-5 The technique illustrated provides excellent insulation value and ensures that your basement will also remain dry and mold free.

Rim Joints

Rim joints are not an obscure arthritic condition, but rather an obscure area of your home where the floor joists meet with the exterior support walls. (They are also known as rim joists or header joints.) Not very exciting true, but they are known to be notoriously difficult to insulate and vapor barrier correctly. Many insulation contractors give only passing effort to this area, stuffing a piece of insulation in place, then stapling a swath of plastic over top, as shown in Figure 3-6a.

A much better method of insulating this area is to use a spray foam material. Although more expensive than traditional methods, this process actually works. As an added benefit, the foam also provides its own vapor barrier, as detailed in Figure 3-6b.

The photograph on the left (Figure 3-6a) details a too typical "stuff and run" rim joist insulation job. The photograph on the right (Figure 3-6b) shows the best way to complete rim joist insulation, using sprayed urethane foam, which provides an integral vapor barrier.

Exterior Wall

Typical framed construction now utilizes 2 x 6 inch (50 x 150 mm) stud walls. This wall cavity thickness is suitable in all construction areas provided quality workmanship is assured. The major problem with exterior wall construction is air leaks. (Have you ever felt the breeze blowing from the electrical boxes of older homes?) The use of blown in fiberglass or cellulose in the wall cavity will ensure complete coverage over the entire surface area. This technique also ensures that insulation works its way around electrical boxes and other obstructions in the cavity. Make sure the contractor fills the wall with a sufficient density of insulation. The material should be tightly packed and not sagging at the top of the frame wall.

The next areas to consider are electrical, phone and cable boxes, which penetrate the exterior structure. Ensure that airtight boxes are used and caulk between them with a 6-mil layer of polyethylene vapor barrier. Ensure the vapor barrier is overlapped and caulked tightly at all seams and with the top and bottom plates making up the wall section.

Air can easily blow between the top and bottom plates of the wall and the floor components. For this reason, these areas should have foam gaskets or caulking compound to ensure air tightness.

Windows

It would be quite easy to write a book on the design and treatment of windows. So, for our purposes, we will only consider them here from an energy standpoint. A single pane of glass is a very poor insulator. During the last 30 years, there have been many advances in window design including double and triple pane versions, low-emissivity coatings, argon and krypton gas fillings and a bewildering array of styles and construction. At the risk of over-simplifying this issue, follow these basic rules when choosing your windows:

- Purchase the best quality window you can afford. Ensure that it is from a reputable supplier and has a good warranty. Ask for references.

- Purchase double-pane windows at a minimum with low-emissivity glass (also known as low-e). This type of glass reduces heat gain and loss. Note that low-e glass and coatings should not be used with south facing windows where winter solar gain is desired.

- Ensure the cavity between windows uses argon gas, or better still, krypton gas. Triple-glazed windows using low-e glass and filled with krypton gas have an insulation value 3.5 times better than traditional double-pane windows.

Figure 3-7 Solar blinds such as these sunshades block 97% of the daylight heat energy and glare while allowing you to see outside (www.shade-o-matic.com). Quilted block-out curtains or bamboo shading blinds will also work.

- All framing materials used in the fabrication of the window should be of wood or vinyl cladding to prevent heat transfer through a metallic structure.

- When installing the windows be sure to adequately insulate and caulk, preferably with spray foam, between the window and house frame

- Make certain that windows are shaded with roof overhangs, awnings or leaves of deciduous trees from early May to mid-October, to prevent unnecessary solar heating. Likewise, ensure that the windows are free to capture the winter sun for the remainder of the year.

Attic and Ceiling Treatment

Provided the attic is not finished, the best way to insulate this area is to blow in 18 inches of cellulose or fiberglass. To ensure that the area above the insulation remains as cool and dry as possible during the year, install adequate roof and soffit vents. Adequately installed vents results in proper air circulation: warm air rises and exits through the roof peak vents while cooler outside air is drawn in through the soffit vents. This upgrade will eliminate moisture damage and ice damming in winter.

If you live in an area where summer air conditioning load is a higher concern than winter heating, you may wish to consider radiant insulation. This material is similar to aluminum foil stapled to the underside of your roof rafters. Working like a mirror, the film reflects heat back through the roof surface, before it gets a chance to hit the attic insulation.

The ceiling vapor barrier should continue down the wall surface a few inches and overlap the one on the wall. Where these overlap, ensure adequate caulking is applied. The best vapor barrier job is wasted if the finishing trades rip, cut or damage the sealed home. We are striving to have you live inside a plastic bag, so take extra care to seal and repair any cuts. Good workmanship will guarantee an energy-efficient home.

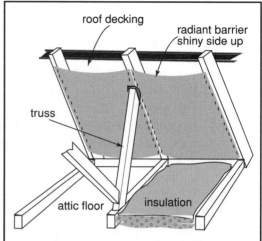

Figure 3-8 Summer air conditioning loads can be reduced by the use of radiant insulation, which reflects heat back through the roof.

Lastly, gasket and seal the attic hatch door in the closed position. This reduces air leakage around its perimeter.

If your home has more complex design features, be sure to discuss these general guidelines with your architect during the planning stage.

Ventilation Systems

Living in an energy efficient house is like living in a plastic bag. With all of the air leaks sealed and the vapor barrier extending from the basement floor to the ceiling, we really have created a sealed, airtight home. Contrast this with the typical house of yesteryear where kitchen and bathroom fans, plus cracks and leaks in the building structure provided fresh air. While these leaks provide ventilation, they are completely uncontrolled and cause tremendous losses in heating and cooling energy. In a typical 30-year-old house, it is estimated that collectively these leaks equate to a 1 square foot hole in the wall!

To ventilate our sealed home, we must resort to alternative measures. The methods you use will depend to a large extent on where you live and whether or not you are connected to the electrical grid.

Most grid-connected, high-efficiency homes utilize a device known as an air-to-air exchanger or heat recovery ventilator (HRV). The HRV is a marvel of simplicity and brilliant in its execution. Stale and moist air from the kitchen, bathrooms and other areas is drawn into the HRV unit and passed over a membrane before being exhausted from the house. At the same time, colder incoming fresh air is pulled into the unit and passed over the opposite side of the membrane. This

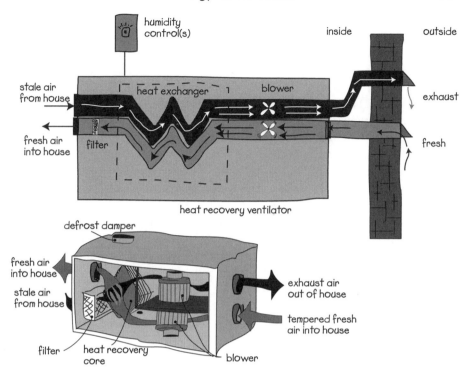

Figure 3-9 The Heat Recovery Ventilator controls air flow into and out of a house. Warm exhaust air is passed over a membrane, transferring this heat energy to the colder, incoming fresh air.

causes waste heat to be transferred to the colder, dry incoming air, which is then warmed and distributed in the central furnace air duct.

Controls and timers located throughout the house can be programmed to monitor humidity and smoke. Even carbon dioxide levels may be monitored, indicating that the house is occupied. Based on rules programmed into the control system, the HRV automatically adjusts air exchange flow to current conditions, while at the same time saving plenty of heating dollars.

Homes in areas that have a low heating load in winter do not fully benefit from the heat recovery aspect of the HRV. However, these homes would benefit from the controlled air intake location and filtration for dust and pollen aspects of the unit.

If there is a disadvantage to HRV units, it is that their fan and electrical circuits cause difficulties in off-grid homes. Although the blower and control circuits may appear to be fairly small loads, in actual fact, they consume a relatively large amount of electrical energy. A typical unit operating 12 hours per day can consume 2,000 W hours of energy.

Unfortunately, there are only a few options that can be used as a substitute for a standard HRV:

- Remove the 120 V fans in the HRV unit and replace with lower flow, high-efficiency 15 W computer style fans. Two 15 W fans operating for 10 hours per day lowers HRV energy consumption to 300 W hours, which is manageable. This modification requires someone who is handy with wire cutters and duct tape.
- Install an air loop intake in conjunction with a bathroom style vent fan.
- Install an air loop intake in conjunction with non-direct venting gas, wood or a pellet stove fireplace.

A review of Figure 3-10 will demonstrate an alternative installation to the HRV that requires no electrical energy. A fresh-air intake is placed in the basement wall near grade level and at least 6 feet (2 metres) away from any exhaust gas vents. An insulated, flexible pipe is connected to the intake, looped and placed just above the basement floor. A damper is installed in the pipe. Typically, the vent outlet is connected in a wall cavity with the outlet facing into a mechanical or laundry room. Such an arrangement limits drafts and ensures that cold air will not flow into the house, until required.

When a kitchen or bathroom vent fan is activated, or a wood stove draws in room air to fuel the fire, stale air inside the house is drawn outside. This creates a partial vacuum that causes fresh, dry, colder air to be drawn into the house. If the intake is placed in a little used area, the fresh air has time to mix with warmer room air before contacting the living area. Although this system is not as efficient as an HRV, intake airflow is controlled. This is a viable alternative to relying on building leakage to make up fresh air, causing uncomfortable drafts. Adjusting the intake damper controls building humidity due to the influx of dryer, winter air.

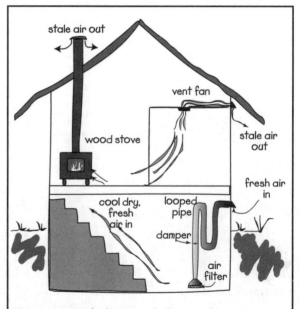

Figure 3-10 A fresh air intake loop used in conjunction with a ventilation fan or woodstove, creates a partial vacuum to exhaust stale and draw in fresh air.

It is important to discuss this or any airflow plan with your local building official to ensure they are onside with the chosen design. It is also important to make this person aware that an off-grid electrical system should not be used to power an HRV unit.

Heating Controls

Have you heard the old story about conserving energy by turning your thermostat down when going to bed at night then turning it back up in the morning? It is a story because no one really does it, at least not for long anyway. Besides, who wants to wake up to a cold house?

Add a little technological marvel to the furnace control: a setback thermostat will take care of reducing the furnace or air conditioning output while you are sleeping and at work. Most models even know the difference between weekdays and weekends, providing you with lots of customization and saving a considerable amount of energy.

New Construction Summary

By following the tips provided above, you can achieve a satisfactory balance between capital cost and energy conservation. In contrast to a home built to standard building codes, a home designed using the principles provided in this (and the next) section allows for a possible reduction in heating and cooling energy requirements and costs by 50% or more.

Updating an Older Home

If your home is more than 10 years old, it is likely that many of the energy-saving features described in the previous section are not incorporated into its design. Don't despair; there is no need to tear down the old place or move. Instead, we can review the entire home systematically and determine where to best put our renovation dollars to make energy sense.

Where to Start?

Every home is unique and, whether it's 10 or 100 years old, what works for one house may not work for the next. As with any project, an assessment is the place to start. We need to check the general condition of the house and test what areas need to be beefed up. Over the last few decades, home designs have improved and architects, engineers and building contractors have all learned from past practices. With this in mind, there is a general checklist we can use to determine what energy shape our house is in right now.

Step 1 - Assessment

- Air leakage is the most common problem with all older homes. Poorly fitting window frames and leaks between doors and chimneys abound. Simple, corrective actions such as applying sealants and weather stripping stop breezes from blowing inside the house. Correcting air leakage also helps regulate hu-

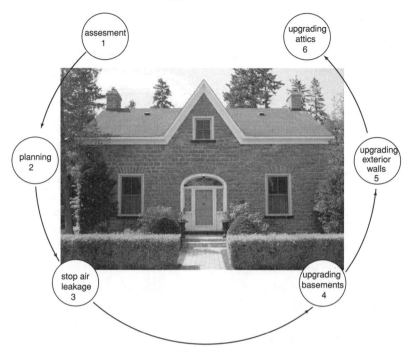

Figure 3-11 Older homes have incredible character and appeal, what they lack is energy efficiency. Proper assessments, planning and retrofit work will always make energy sense.

midity and reduce condensation problems in existing insulation.

- Many older homes treat the basement like an outcast relative: it's there, but leave it alone. Often dark, damp holes, the basement also eats an enormous amount of our heating dollars if not properly insulated. Moisture problems tend to be left to a dehumidifier, which is about as useful as putting a bucket under a leaky roof. Let's try to correct these issues with damp proofing, moisture barriers, drain gutters and proper insulation systems.

- Cavities between walls in older homes are woefully under-insulated or completely uninsulated. Stone walls look great on the outside, but are murder on the heating bill if not properly dressed inside. There are many ways of tackling these problems, from both the inside as well as outside of the home.

- Everyone knows that heated air rises. Let's try to stop it before it escapes through under-insulated attics, leaky joints and chimney and plumbing areas.

Air Leakage

Air leakage control is the most important step that can be taken in upgrading any home. Controlling air leakage provides many benefits including:

- Reduced heating/cooling costs due to infiltration of outside air.
- Increased insulation efficiency; still air is necessary to ensure insulation works.

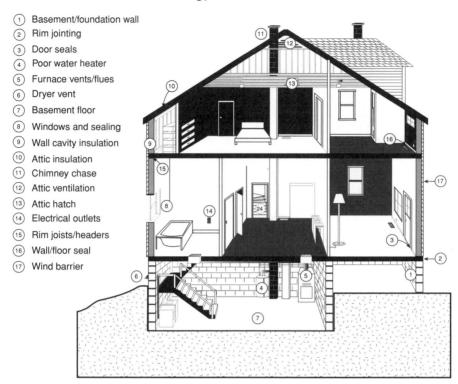

1. Basement/foundation wall
2. Rim jointing
3. Door seals
4. Poor water heater
5. Furnace vents/flues
6. Dryer vent
7. Basement floor
8. Windows and sealing
9. Wall cavity insulation
10. Attic insulation
11. Chimney chase
12. Attic ventilation
13. Attic hatch
14. Electrical outlets
15. Rim joists/headers
16. Wall/floor seal
17. Wind barrier

Figure 3-12 A thousand little air leaks and a poorly insulated area adds up to big energy costs. A thorough assessment of these areas will let you know where you are literally wasting your money.

• Humidity and condensation levels can be controlled.
• Home comfort is improved as drafts and cold spots are eliminated.

It is possible to hire professional contractors to assess the quality of the air barrier system in your house. Alternatively, a simple yet effective method is to conduct the test yourself using several incense sticks left over from your psychedelic days in college.

Hold the incense near the suspected leakage area on a windy day and observe what happens to the smoke. Smoke drawn towards or blown away from a suspect area is demonstrating an air leakage path. Mark the location down on your list. Continue testing in this manner for:

• electrical outlets, including switches and light fixtures
• plumbing penetrations that include the attic vent stack, plumbing lines to taps, dishwashers, etc.
• floor-to-ceiling joints and other building areas that are storied; around baseboards, crown molding and doorway molding.
• fireplace damper area and chimney exiting through the attic or wall

- attic hatch (ensure that it is properly sealed)
- windows and doors (check them thoroughly to ensure the glass fits tightly and that the casing area is sealed)
- appliances' ventilation (kitchen fans over the stove, bathroom fan pipes, dryer vents, gas stoves, water heaters, etc.)
- pipes, vents, wiring and plumbing lines in the basement and attic (In the attic, move the insulation if necessary to access these areas.)

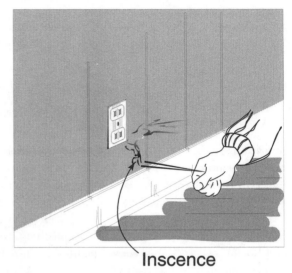

Inscence

Figure 3-13 A burning incense stick held near suspected areas of leakage on a windy day would cause the smoke to move towards or away from the leak.

In an older home, it is quite likely that the smoke from the incense will blow just standing in the middle of the room. Don't despair, just mark all of the areas down on your sheet and perhaps record a "severity rating" from 1 to 10. This way, you can tackle the tough problem areas first.

The Basement

Most homeowners don't even consider unfinished basements as a source of heat loss. Part of this mentality comes from the mistaken idea that heat rises and that earth is a good insulator. Both are wrong. Warm air rises because it is less dense compared to cool air. Heat travels in any direction it chooses, but it always travels from a warm area to a colder one. Additionally, older basements have large surface areas of uninsulated walls and flooring that act as a heat sink.

As we discussed in *Air Leakage*, there is also a lot of leakage through crevices in the walls, windows and at the top of the foundation wall where it meets the first floor. An uninsulated basement can account for up to 35% of the total heat loss in a home.

It is not possible to simply add insulation and air barriers to a damp, leaky basement without first correcting any underlying problems. Any areas that accumulate water in the spring or after a heavy rain must be repaired as wet insulation has no energy efficiency value and will ultimately contribute to mold and air quality issues. Check the basement for dampness, water leaks and puddles in wet times, and major or moving cracks in the foundation wall. Also ensure that a sump pit and pump have been installed in areas where persistent water accumulation is present.

Some basement wetness problems are corrected by sloping the landscaping away from the foundation wall, or adding rain gutters to the house eves. If you cannot ensure that these measures will correct the basement water problems, then it will be necessary to excavate around the perimeter of the house to provide supplementary drainage and to apply damp proofing on the foundation wall as well as external, waterproof insulation treatments.

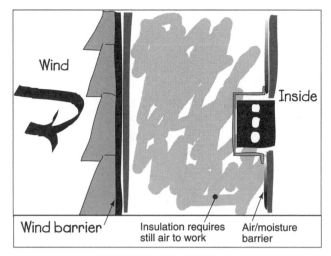

Figure 3-14 Proper wind and air barriers combined with adequate insulation keeps the heat in. If any one part is missing or faulty, you may be throwing money away.

Exterior Walls

Exterior walls account for a large percentage of the heat loss in the home. In addition, high wind pressure increases air leakage, further contributing to heat loss. Older homes were built in a large variety of construction techniques, some easier to insulate, seal and vapor barrier than others.

Wood Framed Walls

Wood framed walls are the easiest to insulate, as they are very similar to new construction. The major concerns relate to wall cavity thickness and access. When assessing these wall structures, attempt to determine if the wall is empty or if there is already some form of insulation in place.

The simplest means of checking these walls for insulation is to remove the cover plate of an exterior wall electrical plug or switch and using a flashlight and a thin probe check for insulation behind the electrical box. CAUTION! Before completing this test, ensure that the circuit breaker or fuse is removed from that circuit and test the outlet or switch to verify that it is off. This test will have to be conducted at a few points around the house to ensure that your sample investigation is accurate.

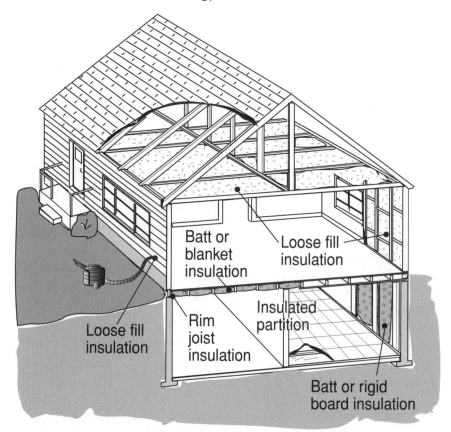

Figure 3-15 The older home can be upgraded by adding insulation to the structure as shown above. Supplementary insulation as well as air and vapor barrier may also be added.

Brick Walls

Brick walls on homes are almost always constructed of veneer with a frame wall on the interior side. Oftentimes there is a small air gap between the brick and the frame wall to allow air to circulate and prevent moisture on the brick from rotting the framing members.

This air gap must not be filled with insulation. However, the frame wall might allow for insulation to be added. Use the same tests as those used for a standard frame wall to determine the depth and area that may be insulated.

Stone or Other Solid Walls

Stone, concrete block, cut stone and other solid wall treatments are similar in nature to brick walls, with or without a framed wall. Generally, these walls are not suitable for insulating and sealing and require extensive reframing on the inside or special siding and coverings added to the exterior.

Attics and Roof Areas

Most homeowners love to dump their home insulation upgrade dollars in the attic. Perhaps it's the mistaken idea that heat rises and the losses must therefore be highest in the attic? Could it also be because attics are one of the easiest places to insulate first? Just dump a few bags of insulation in the old attic and, *voilá!,* your heating bill goes down by 50%?

Sorry, not so fast. The attic does lose heat, true, but it actually loses less than an uninsulated basement or exterior wall. Most homeowners, no matter how old their house, may have had a passing thought about attic insulation and may have even added a few inches of something up there. Adding more insulation on top of old is not a problem, except if the air leakage tests completed earlier confirmed that problems exist. If this is the case, it may be necessary to remove or move aside existing insulation to gain access to the leaking area. The necessity of ensuring quality air sealing in the attic area cannot be understated.

Another problem area in the attic is moisture. This moisture may originate from different sources such as a leaky roof, ice damming in winter or frost. Moisture in the attic can also come from within the house through leaky ventilation fan outlets coming from the bathroom and kitchen areas.

Ventilation of the attic itself is important to provide summer cooling and winter dryness. Many older homes do not have sufficient (or any) vents in the gable, soffit or roof peak. Check that these are present. The standard vent area (for both intake and outlet vents) should be 1 for every 300 units of attic floor area. For example, a bungalow having a perimeter of 24 x 50 feet would have an attic floor area of 1,200 square feet. Dividing the attic floor area of 1,200 square feet by the ratio of 1 to 300 gives us a total required vent area of 3 square feet.

A common practice in upgrading is to add *electric* vent fans into the attic area, usually in the gable or roof peak. This is not required, nor is it recommended. An electric exhaust fan increases airflow in the attic and may exceed the intake capacity of the soffit intake vents. If this should occur, additional air will be drawn from the main part of the house, which is exactly the opposite of what is desired.

Check the attic several times during the year, such as after a heavy rain or very cold day. Look for wet areas, mold, rot or small "drip holes" on the insulation or attic floor surfaces.

Attics come in all shapes and sizes. If yours is stuffed to the rafters with a lifetime of family heirlooms, maybe you already have enough insulation. For others, the attic may form part of the living space or be replaced with cathedral ceilings. The trickier it is getting access to the attic, the more difficult upgrading the insulation system will be, and therefore professional assistance may be required.

Step 2 – Planning The Work

As this is a step-by-step handbook, you may wish to tackle some of the upgrading work yourself. Most of the tools required are pretty common household items and the few specialized tools can be rented from your local rental depot or borrowed from a friend. One couple recently completed a blown-in cellulose cathedral ceiling while balancing on a scaffold and using an insulation blower loaned to them by the material supplier. The do-it-yourself approach might not be for everyone, but if you do it correctly, it can result in a considerable cost savings.

Some upgrading work is best left to the professionals: urethane foam spray insulation, for example, requires a truckload of specialized equipment; and if excavating around the foundation of the house is a job that just doesn't make the top of your "list of things to do in life", a backhoe and a experienced operator really work wonders here.

Building Codes
The various national, state and local building codes and their variations are enough to frustrate any do-it-yourselfer. But don't cut corners. The reason the codes are there in the first place is for your health and well-being. Get to know your local building official. Although there are plenty of horror stories circulating about these officials, they are generally started by people who began the upgrade work before completing the planning and getting a building permit to begin with. You will find that most officials are very helpful, providing guidance on technical issues and referrals to qualified contractors or suppliers in your area. Work *with* them.

Safety
Working around the house, climbing ladders, working with insulation and chemicals can be dangerous. Make sure you have the proper safety equipment including work boots, dust masks, rubber or latex gloves and eye protection. Attic and basement areas may not be well illuminated; ensure that you have a suitable light. Use caution on the attic "floor". Often this floor is nothing more than the drywall or lath and plaster finish material on the ceiling above. It will barely hold a cat, let alone your body weight.

Step 3 – Stopping Air Leakage
Now that we are armed with our "list of air leaks" from the assessment phase, we can start getting down to business:
Caulking
Seal up small cracks, leaks and penetrations on the inside (warm side) of exterior walls, ceilings and floors. Sealant applied on the inside lasts longer as the material is not exposed to the elements outside.

Caulking is done using an inexpensive gun with a tube of appropriate material. There are literally hundreds of caulking materials available. Discuss the type best suited to your project with a building supply store. After a caulking job, many people are dissatisfied with the caulking brand they used and/or the job they did. Avoid the tendency to purchase poor quality materials which are difficult to apply and do not last.

Remember to purchase high temperature silicone or polysulfide compounds for areas around wood stove chimneys or hot water flue vents.

Figure 3-16 It takes a bit of practice to make a good caulking joint. Go slowly and don't cut the nozzle too large, otherwise you'll end up with hard-to-clean goop!

- Locate the area to be caulked. Determine the compound type appropriate for the job.
- Never caulk in cold weather. Caulking should be applied as close to room temperature as possible.
- Clean the area to be caulked. Large cracks and holes greater than 1/4 inch will require a filler of oakum or foam rope sold for this purpose.
- The nozzle of the compound should be cut just large enough to overlap the crack. Insert a piece of coat hanger wire or long nail into the nozzle to break the thin metal seal.
- Pull the caulking gun along at right angles to the crack, ensuring that sufficient compound is dispensed to cover both sides of the crack. Remember that caulking shrinks, so it's better to go a bit overboard.
- The finished caulking "bead" should be smooth and clean. The surface of the bead may be smoothed with a finger dipped in water. Some compounds require paint thinner or other chemicals to clean up. Check the label of the compound before using it to determine the cleanup method.

Electrical Boxes

Air leakage around electrical boxes is so common that there is an off-the-shelf solution available. Special fireproof foam gasket and pads may be added just behind the decorative plate as shown in Figure 3-17. To ensure a superior seal, use an indoor latex caulking compound on the gasket face before applying the decorative cover.

If the room is being renovated at the same time, install plastic "hats" around the entire electrical box. These hats have an opening for the supply wires and small flaps that can be sealed to the vapor barrier. Ensure the area where the supply wire enters the hat is well sealed with acoustic sealant.

Windows

Older homes often have single panes of glass puttied into wooden frames. When the putty dries out, air leakage occurs. Remove old putty and replace it with glazing compound to ensure a high degree of flexibility and a long life. Old-fashioned putty is not recommended.

The area between the window and the frame is another area prone to leakage. Access to this area usually requires the removal of the window casing trim. Use oakum, plastic foam rope or, better still, urethane foam spray to air/vapor seal and insulate this area in one application.

Baseboards, Moldings and Doorway Trim

Trim pieces such as baseboards, moldings and door casing are used to cover the gap between one framing section and another. For example, a pre-manufactured door is placed in a framed section of the wall called the "rough opening". Obviously the door must be smaller than the opening to fit. If this opening is not properly sealed, considerable air leakage will result.

The best way to seal these areas is to use methods similar to those described for window framing sections earlier in this chapter, and as is illustrated in Figure 3-18. For smaller gaps or areas where the urethane foam may be too messy (near carpets or finished wood), Oakum or plastic foam rope may be jammed into the gap.

Fireplaces

Fireplaces warm the heart, but rob you blind. Air leakage up and down the chimney when the unit is not in use is enormous. When the fireplace is going, the

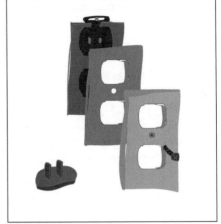

Figure 3-17 Air leakage is so common in older homes that off-the-shelf sealing pads have been designed to solve this problem.

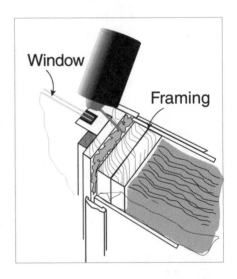

Figure 3-18 Urethane foam spray works wonders in small nooks and crannies like between the window and house framing.

suction created by the roaring fire draws large volumes of expensive heated room air up the chimney. This suction draws in cold outside air to replace the heated air that literally went up in smoke!

What to do? Simple. Replace the fireplace with an airtight wood burning stove or similar controlled combustion unit. We will discuss these items in greater detail in Chapter 4.

If you really must keep the fireplace, make a removable flue plug that properly seals the chimney when not in use. Check to make sure the damper closes as it should. If you detected air leakage around the chimney and framing as part of your assessment in Step 1, seal any cracks with heat-resistant sealant and mineral wool or fiberglass batting.

A quick word on glass fireplace doors; don't waste your money. If you like the look of them or are worried about flying sparks, so be it. Be warned that the majority of these door units are cheaply made and will not provide any air sealing capacity.

Attic Hatch

Seal the attic hatch in the same manner as your front door. Use hook and eye screws to ensure that the hatch stays well sealed against the weather strip material as detailed in Figure 3-19.

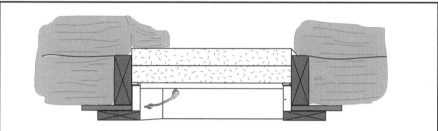

Figure 3-19 The attic hatch is often forgotten when it comes to air leakage and insulation.

Air and Vapor Barrier

So far in our discussion on air leakage, we have been concerned about the worst but most easily corrected areas. If you are undertaking more extensive renovations, it may be possible to improve large sections of the air and vapor barrier. Before we continue with how to do this type of major upgrade, let's quickly review the function of each component to be sure we understand how this work can be implemented Figure 3-14 details the function of each component.

On the exterior of the house, directly under the siding material, is a layer of spun-bonded olefin, which is often sold under the trade name Tyvek®. This material stops wind from penetrating into the insulation area and repels rainwater. It is also permeable to water vapor, which allows trapped moisture in the insulation cavity to escape outdoors, preventing rot.

On the warm side of the insulation is the vapor barrier. In newer homes, this barrier is fabricated with 0.006 inch (6mil) thick polyethylene sheet (think food wrap on steroids). Large sheets of vapor barrier are affixed to the wall studs and carefully sealed during construction.

The vapor barrier acts as a secondary wind barrier, stilling the air trapped in the insulation. It also prevents warm, moist household air from penetrating into the insulation (remember, heat moves toward colder areas). Moist air contacting cold insulation and wood siding on the exterior side of the insulation will quickly condense into water. Should this water be present for extended periods of time, it will cause structural rot and mold problems. Many a house has been seriously damaged due to high inside humidity coupled with an insufficient vapor barrier.

Your Options

During renovation, it may be possible to install both an air and vapor barrier. This type of upgrade can only be completed when the framing members are accessible, which typically occurs when a major rebuilding effort is in progress or an addition is being constructed. Major renovation involving more than just a few sheets of vapor barrier is discussed in *Step 5 – Upgrading Walls*. It doesn't matter if your renovation work is concentrated on the exterior or interior side of the wall, there are still ways to complete the work.

Exterior Wall Upgrade

If you are upgrading, removing or adding on top of existing siding, it is possible to increase the air barrier with little difficulty. Removal of the old siding reveals the wood sheeting covering the framing members. A layer of Tyvek® air barrier can be taped to the exposed wall sheeting or over existing smooth siding materials. The air barrier is similar to the vapor barrier in that all joints should be carefully taped to ensure a continuous barrier to wind penetration.

In this example, the interior wall is not being upgraded. Follow *Step 3 – Stopping Air Leakage*.

Figure 3-20 A home wrapped with a suitable air barrier decreases air leakage and increases insulation values.

Although a polyethylene film is not being applied, a reasonable quality vapor barrier can be made using multiple layers of latex paint over well-sealed drywall. When an external air barrier is mixed with air leakage sealing and latex paint application, you can achieve the next best thing to new construction.

Interior Wall Upgrade

Your choices for upgrading air and vapor barriers increase if you are updating an interior wall. We will discuss this work further in *Step 5 – Upgrading Walls*.

Step 4 – Upgrading Basements

In *Step 1 – Assessments for Basements*, we discussed the need to ensure complete dryness before insulation and/or vapor barriers can be added to the inside. Persistent moisture or water leakage problems will ruin even the best quality work.

Dampness

Minor dampness causes staining or mold growth, blistering and peeling of paint, moldy smells and efflorescence or whitish deposits on concrete. Provided dampness problems are limited to the above, they may be corrected from the inside by the cleaning and application of damp proofing to the

Figure 3-21 Properly installed rain gutters and sloping the grade away from the house are two simple, yet effective means of keeping the basement dry.

foundation wall. More serious problems will have to be corrected from the outside.

Major Cracks

If your foundation wall has large cracks or ones that appear to be getting larger, prior to upgrading, seek professional help to determine if structural repairs are necessary.

What Are My Choices?

There is no doubt that insulating from the inside of the basement is the easiest and lowest cost, provided the basement is dry. Insulating from the outside will do the best job from a technical point of view. What to do? Let's begin by weighing the pros and cons of each method.

Insulating Inside

Insulating inside usually involves the addition of a wood framed wall and adding some form of insulation material, similar to a standard wall. The advantages of insulating inside are:

- The work can be done at any time of the year.
- A completed job will increase the value of your home.

Insulation	Type	R-Value (approx.)	R.S.I Value
Batts	Fiberglass or Rock Wool	3 1/2" (R-11) 5 1/2" (R-19) 9 1/2" (R-30)	90mm 1.9 140mm 3.3 240mm 5.0
Loose Fill	Fiberglass or Rock Wool	R-2.7 per inch	1.9 per 100mm
	Cellulose	R-3.7 per inch	2.6 per 100mm
Rigid Board	Expanded Polystrene (Beadboard)	R-4 per inch	2.8 per 100mm
	Extruded Polystrene	R-5 per inch	3.5 per 100mm
	Polyurethane or Polyisocyanurate	R-7 to R-8 per inch	4.4 - 5.6 per 100mm
	Polyurethane	R-7 to R-8 per inch	4.9 - 5.6 per 100mm

Table 3-2 Common Insulation Materials and their Heat Resistance Values in R/RSI.

- Indoor finished space will be increased.
- It is the lowest cost way to update basement insulation and vapor barrier.
- The landscaping, porches, walkways and other obstructions outside will not be disturbed.

The disadvantages to insulating inside the home are as follows:

- Persistently damp or wet basements cannot be upgraded
- Furnaces, electrical panels, vent pipes and plumbing obstructions make do-it-yourself framing more difficult.

Insulating Outside

Insulating outside the home involves more complex and expensive excavation but at the same time offers the ability to correct latent foundation defects. The advantages of insulating outside are:

- The outside wall is generally straighter and simpler to insulate once the excavation has been completed.
- Most moisture and water leakage problems can be corrected at the same time.
- Foundation cracking and other damage can be inspected and repaired.
- There is no lost space inside the home due to the added wall thickness.
- The weight or mass of the foundation is on the warm side. This mass absorbs heating and cooling energy, helping to balance temperature fluctuations.

The disadvantages of insulating outside the home are:

- Excavation work is costly and may be difficult if there are porches, finished walkways or decks abutting the foundation wall.
- Storing the excavated dirt will damage lawns and bring mud and sand into the house.
- Work cannot be done (economically) in the winter season.

The next step after weighing which system works best is to examine how the work is done.

How to Insulate Outside

Step 4 - 1 - Preparation

- Prior to beginning the insulation work, remove any outside features that will get in the way. This includes decks, stairs, trees, walkways and so forth.
- Determine where power, water, sewer, septic tank lines, gas, telephone and other services enter the building. Contact your utility to locate unknown pipes and wires; this is usually a free service.
- Determine where the excavated dirt will go. Placing a polyethylene sheet or tarp on the grass to hold the dirt will make the clean up easier.

Step 4 - 2 – Excavation

- USE CAUTION! The soil may be unstable and fall back into the excavation causing injury or death.
- The excavation must extend down to the top footing
- Never dig below or near the base of the footing (see Figure 3-22, item 8). This will cause the foot to sink and the house to drop.
- USE CAUTION! Older rubble stonewalls may collapse without the support of the surrounding soil. Seek expert help if you are in doubt.

Step 4 - 3 – Preparing the Foundation Wall

- Brush and fully clean the foundation wall. Scrape any loose concrete or rubblework from the wall (Item 2)

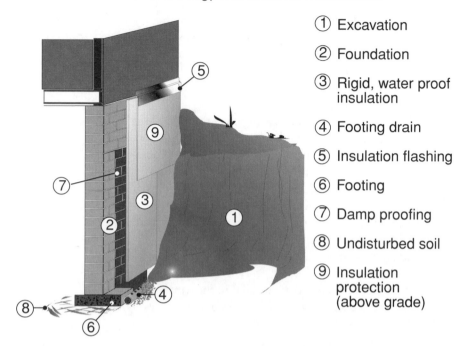

① Excavation

② Foundation

③ Rigid, water proof insulation

④ Footing drain

⑤ Insulation flashing

⑥ Footing

⑦ Damp proofing

⑧ Undisturbed soil

⑨ Insulation protection (above grade)

Figure 3-22 Excavation of the outside foundation wall is the best way to solve water leakage and foundation structural problems while upgrading insulation.

- If concrete is missing or damaged in places, it is best to apply a coating of parging (a waterproof type of masonry cement) to the damaged or missing sections. Allow this to dry.
- Apply damp proofing compound to the foundation wall (Item 7). Damp proofing is a tar like substance that is painted on the foundation wall from the top of the footing to grade level.
- Inspect the footing drain system (Item 4). If there is no footing drain, determine whether one can be added:
- The footing drain is made by installing a flexible pipe that has been manufactured with thousands of holes along its length.
- The pipe is normally supplied "socked" (with a pantyhose like liner to prevent sand and dirt from entering).
- This pipe is laid down adjacent to the footing, without affecting the undisturbed soil in this area.
- A 12 inch layer of "clear stone" is spread on top of the footing drain.
- A strip of filter fabric or "gardeners' cloth" should be applied on top of the clear stone, to prevent sand and dirt from plugging the footing drain.

- The footing drain should be routed downhill, away from the house to a drainage ditch or dry well. A dry well can be simply made by excavating a 3 to 4 foot diameter hole to a depth lower than the house footings. Place the end of the footing drain pipe in this hole and fill with crushed "clear stone" to within 12 inches of the top. Cover the top of the dry well and supply ditch with soil to finish.

- Check that all services penetrating the foundation wall are well sealed with a suitable caulking compound. Remember to allow room for the insulation.

Step 4 - 4 – Installing the Insulation

- There are many types of insulation in use today, but the most common and easiest to work with is polystyrene rigid board. This material is supplied in sheets that are typically 2 x 8 feet long and have interlocking grooves running the full length. As indicated in Table 3-1, the minimum recommended insulation level of rigid board is R13/RSI 2.0. Your building supply dealer will be able to recommend which brand of insulation is supplied in your area, just make sure you explain your R/RSI rating requirements.

- Rigid board insulation is very fragile and will break when exposed to wind or undue flexing. Make sure the board is protected prior to and during the installation phase of the work. A special flashing trim is installed at the top of the foundation wall that clips the insulation board in place. Alternatively, pressure treated plywood may be applied on-site to hold the insulation to the header joist at the top of the foundation wall. (Item 5).

- Make sure that all insulation joints are well sealed and clipped together. It is important that you discuss your specific needs for a flashing and insulation clip system with your material supplier as there are many variations available.

- Ensure insulation overlaps at the wall corners, preventing areas of concrete from being exposed.

- A covering is required to protect the insulation from sunlight and damage from traffic and animals where it protrudes above grade (Item 9). Coverings may be purchased for the application or fabricated from any of the following:
  - pressure-treated plywood
  - vinyl or aluminum siding to match the house
  - metal lath and cement parging

Step 4 - 5 – Backfilling the Excavation

- After backfilling the drain tiles as discussed in Step 3, it is time to refill the excavation. If the soil that was removed earlier is heavy clay or drains poorly, it is recommended to remove it and use clear-running "pit run" sand as the back fill material. This will greatly assist in encouraging water to drain away from the foundation wall, further ensuring a dry basement.

- When the excavation is refilled, ensure the finished grade slopes away from the house. This promotes drainage and allows runoff to move away from the foundation wall. This is a good time to remind you to install eaves troughs with downspout pipes that lead away from the house.

How to Insulate Inside the Basement

Insulating a basement that is known to be dry is not much different from insulating a new house. The major differences relate to the type of wall structure that is used. A fairly new poured concrete or block wall should present few problems. Older rubble or cut stonewalls tend to be "wavy" and vary in height somewhat. These issues make framing more difficult, but do not change the methods involved. The dry basement insulation method is just a repeat of new home construction.

* A tarpaper style moisture barrier is glued to the concrete wall starting from just below grade, and then it is folded up and glued to the bottom of the interior vapor barrier as detailed in Figure 3-5.

* The basement wall is fabricated using standard framing techniques, except that the 2 x 4 inch stud wall is "pulled" away from the foundation wall by 4 inches. This increases the wall cavity space for additional insulation without utilizing additional framing material.

* Instead of traditional batt insulation, blown in fiberglass or cellulose is suggested. These materials ensure complete coverage and packing density, especially in the hard-to-insulate areas around plumbing lines, electrical boxes and wiring.

* All other finishing details are done exactly the same as is done in a new home.

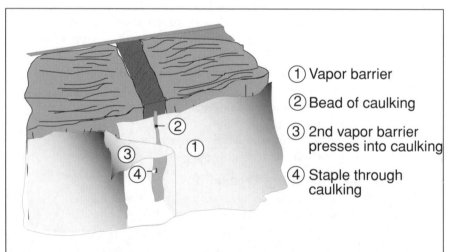

1. Vapor barrier
2. Bead of caulking
3. 2nd vapor barrier presses into caulking
4. Staple through caulking

Figure 3-23 A vapor barrier is installed in large continuous sheets. Where the sheets meet, they must overlap and be well sealed with acoustical sealant .

Problem Basements

There are a number of obstacles that can be encountered in older basements, including:

* Dirt or packed stone floors
* Crawl spaces or basements with very low clearance heights
* No basement, slab-on-grade construction
* Building is on piers or blocks

While it is possible to insulate problem basements it is recommended that you seek a professional contractor to review your specific requirements. There are too many variations in climate and construction type to generalize on how to deal with each scenario.

Step 5 - Upgrading Exterior Walls

Walls account for a sizeable percentage of the heat loss in houses due to the large surface area they cover. Older homes, constructed of solid material such as stone, brick or log, gave little thought to interior insulation. Many of these designs have a small air space between the brick and the small inner frame wall. This area must not be insulated, as it is used as a drainage cavity for water leakage and condensation.

Hollow concrete block walls should not be filled with insulation. The quality of the insulation and the "thermal bridging" effect of heat and cold passing through the block does not warrant the trouble or expense.

Traditional frame walls are more easily insulated as there is usually an accessible cavity. Using various construction techniques, you can determine if there are cables, duct work or other obstructions inside the wall cavity that may interfere with the application of the insulation .

Solid walls of stone, brick or log may be insulated from either the inside or outside, depending on several factors.

* Heritage appeal. Homes built of traditional stone, log or brick may be too beautiful to cover. Insulating from the inside may make the most sense in these situations.
* Outside requires finishing. If your exterior siding or building material is looking a little tired, there are a number of ways to insulate from the outside and refinish the exterior siding at the same time.
* Interior requires new lath and plaster walls or general updating. When the inside wall surface is cracking and the wallpaper is starting to get to you, perhaps insulating from the inside is the right choice.
* A bit of exterior and interior finishing is required? If energy efficiency is the ultimate quest, why not consider doing both? It is possible to add insulation to both the outside and inside exterior walls, creating a "good as new" insulated home.

Construction type, material costs and your skill level will determine the path you take on this upgrade.

Frame Wall Cavities

Frame wall cavities are by far the easiest to upgrade, provided they are empty. A wall that is half-filled with insulation from an earlier job makes it almost impossible to do a good job and ultimately will not be worth the expense. During the assessment stage, we determined how much insulation was in the wall by using a flashlight and poking around through electrical outlets (with the power off!) and other access points.

If there is little or no existing insulation, a contractor will apply either cellulose fiber or polyurethane foam into the cavity. These materials are applied through small holes drilled into either the exterior siding or interior drywall finish.

The holes drilled into exterior siding are plugged using wood dowels. Once the siding is re-painted, the work is almost invisible. Brick homes that are suitable for blown insulation have selected bricks removed. The insulation is sprayed and the bricks are replaced.

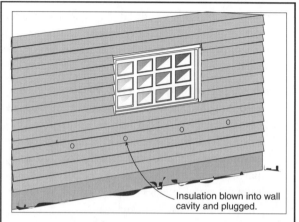

Insulation blown into wall cavity and plugged.

Figure 3-24 Cellulose or urethane foam insulation may be blown into the wall cavity from either the interior or exterior side of the wall.

If the work is being done from the inside, a hole is drilled into the drywall or lath and plaster surface at strategic locations. The holes are filled, primed and re-painted.

If the work is being done inside and the interior drywall or lath work is poor, consider drilling holes into the existing surface, then cover it with a full vapor and air barrier. New drywall can then be placed directly on top of the existing surface. This technique uses up very little space and is cheaper than a fully framed interior wall.

Upgrading Insulation on the Exterior

If it's time to upgrade the old siding, it is also an excellent opportunity to upgrade exterior insulation as well. There are several methods available that allow you to add insulation under the new siding, significantly increasing the overall "R/RSI" value of the home. This work can also be coupled with blown-in insulation, eliminating the need for filling and re-painting access holes. Consider these points as well:

• It is possible to add a generous amount of insulation by using high-density

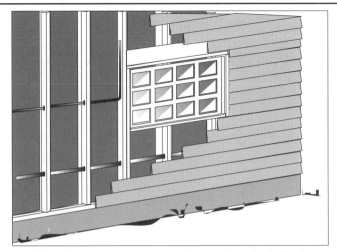

Figure 3-25 When choosing a contractor, make sure they check for obstructions or blockages in the wall, including window and door frames.

rigid insulation sheets or creating a new wall cavity over top of the existing siding.

• If the house is poorly insulated or made of stone, solid brick or masonry, add a vapor barrier directly to the inside top of this surface, under the new insulation. Follow the rules outlined above regarding proper vapor barrier installation and sealing.

• Remember that doors and window openings will have to be extended to allow for the additional insulation thickness.

• Ensure that water runoff from the roof will not drip between the old and new wall. This will ruin the insulation and cause structural rot damage. In this case, an eaves extension or flashing is necessary.

• Make sure that the new insulation is well air-sealed. There is not much sense in having a great insulation job, with the winter winds howling between the old and new work.

How to Do It

There are dozens of ways exterior siding can be applied. We will review the most popular methods and then point you to your local building supply contractor for more detailed information about materials for your specific application. Regardless of the application method you choose, be sure to wrap the entire upgrade work in spun-bonded olefin (Tyvek® brand) air barrier and ensure that it is carefully wrapped and taped at all joints.

1. It is possible to add exterior insulation by simply purchasing insulated siding materials. This is the easiest approach, although the insulation values are somewhat limited owing to the small thickness. Check with your building supplier

on the many finishes available. These pre-insulated siding products install in the same manner as their uninsulated counterparts.

2. Rigid board insulation can be applied directly over existing surfaces using appropriate fasteners and adhesives. Your building supplier or contractor can review which materials will work best in your application. If rigid board insulation is used, make sure that all joints are tight and well taped to reduce air leakage.

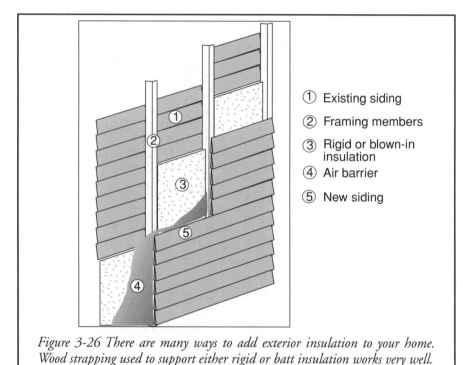

① Existing siding

② Framing members

③ Rigid or blown-in insulation

④ Air barrier

⑤ New siding

Figure 3-26 There are many ways to add exterior insulation to your home. Wood strapping used to support either rigid or batt insulation works very well.

3. Rigid board or batt insulation may be added to a new wall framed on top of the existing siding structure. Batt insulation requires a thicker wall to achieve the same insulation value as rigid materials.
   a. Frame the desired wall thickness on top of the existing wall siding.
   b. Add the insulation, ensuring complete coverage between studs.
   c. Apply an air barrier of spun-bonded olefin, using well-taped edges and seams.
   d. Ensure that water runoff from the roof will not drip directly on the top of the new wall extension.
   e. Make sure that wind cannot enter into the wall cavity from the top or bottom framing plates.

Upgrading Insulation on the Interior

Determine the best application method by assessing your specific requirements. Your choices include:

•   Upgrading an existing wall. This is often done when the existing wall material is damaged or lath and plaster is falling off.

•   Adding rigid board insulation directly to an existing wall surface. Once the wallboard is added, a new drywall finish surface is commonly added.

•   Build a new frame wall. The new wall takes more space from the interior, but provides lovely window boxes. It also allows you to increase home insulation values and proper vapor barrier.

Upgrading an existing wall

A common upgrade for older homes is removing the lath and plaster wall (a messy job at best) and replacing it with modern drywall material.

•   With the internal wall studs exposed, you can easily upgrade wiring, plumbing, ductwork, central vacuum pipes and, of course, your insulation.

•   Use additional horizontal strapping as shown in Figure 3-27 to increase the wall cavity depth from a typical 2 x 4 inches to 2 x 6 inches.

•   Electrical boxes and window frames must be extended to allow for the increased wall cavity depth. Building supply stores carry box extenders for such work.

•   Use batt insulation layered vertically for the existing wall and horizontally for the new wall section. Alternatively, blow in cellulose insulation after the vapor barrier has been installed.

•   Install a 6-mil vapor barrier over the new framing studs.

•   The drywall or surface finish may now be applied.

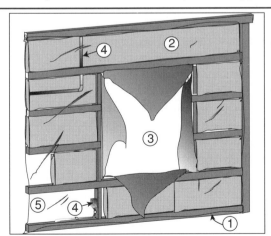

① New framing wall

② Insulation

③ Rough window opening

④ Electrical/plumbing runs

⑤ Vapor barrier

Figure 3-27 Adding a new, non-load bearing wall is very similar to the methods used insulating an inside basement wall. This method should allow you to achieve increased home insulation values.

Add Rigid Board Insulation to an Existing Wall

Rigid board insulation may be applied directly to an existing wall surface.

- As with upgrading an existing wall, windows and doorframes must be extended and electrical box extenders added.
- Your building supply store will be able to provide suitable fasteners to hold the rigid board insulation to the existing wall surface.
- Vapor and air barriers are not required when rigid board insulation is installed, provided the seams are snug and well taped.
- Drywall or other finish materials may be applied directly to the rigid board insulation.

Building a New Frame Wall

You can build a new frame wall by using exactly the same techniques described earlier in *Upgrading an Existing Wall*. You may also wish to review the points outlined in Figure 3-5.

Insulation on Both Sides of an Exterior Wall

A final word is required on adding insulation to both the inside and exterior of the home. If you wish to add some of the insulation on one side of the wall and additional insulation on the other, be careful to limit the ratio to 2/3 inside and 1/3 outside. This mix is required to ensure that condensation does not damage the insulation system during the heating season.

Step 6 – Upgrading Attics

Air Leakage (a quick review)

Earlier, in *Step 3 – Stopping Air Leakage*, we discussed the need to eliminate air leaking into the attic space from the warm areas below. Preventing air leakage stops the warm, moist air from condensing when it reaches colder winter air inside the attic. Condensing moisture may cause mold and structural damage due to rot. In minor cases, moisture buildup greatly reduces insulation effectiveness.

Installing a proper vapor and air barrier requires installation on the warm side of the insulation, NOT ON THE COLD SIDE! From a practical point of view, adding a large polyethylene film to the underside of the insulation is very difficult. It requires moving large amounts of existing insulation and installing the barrier around obstructions such as duct work, plumbing pipes and wiring.

Should the interior ceiling of the house also be under renovation, an opportunity is present to add the air and vapor barrier directly to the attic-framing members, under the drywall.

In the majority of cases, it is just not possible to add an effective air and vapor barrier into a retrofit attic. In these cases, a well-sealed and caulked attic must suffice.

Attic Ventilation (in case you forgot)

Attic ventilation is mandatory. You will recall that a ratio of 1 to 300 of roof ventilation to attic floor area is required (approximately 3 square feet of vent space for

Figure 3-28 Soffit vents form the air intake section of the attic ventilation system.

Figure 3-29 Roof vents are added on the roof peak to vent hot air out. You will require a minimum of 1 square foot of roof venting for every 300 square feet of attic floor space.

every 1,000 square feet of attic floor over a heated area). A home will typically have a row of vents under the eaves called soffit vents shown in Figure 3-28. These vents form the air intake for the attic ventilation system. Roof vents are installed along the roof peak (see Figure 3-29) and allow hot attic air to exit. Air exiting the roof or gable vents causes suction within the attic, drawing in cooler air through the soffit vents.

Before starting any insulation work in the attic, ensure that the vents described above are present and of adequate size. If in doubt, a building contractor can determine their effectiveness.

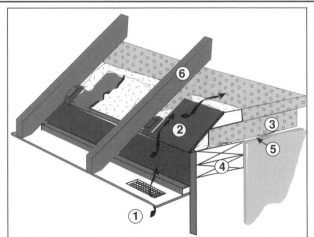

① Soffit vent

② Cardboard baffles

③ Insulation (R-40/RSI 7.1)

④ Top of wall

⑤ Vapor barrier

⑥ Roof truss or rafter

Figure 3-30 A well-sealed attic with plenty of ventilation and insulation are the final steps in keeping your house warm in winter and cool in summer.

Installing Attic Insulation

Once the attic has been well sealed against air leakage and the soffit and roof vents checked, we are ready to begin insulating. You have a number of choices as to which material to use. Rigid board insulation is almost never used, due to the difficulty of working in confined spaces.

Blown or poured fiberglass and cellulose are perhaps the easiest insulation materials to use. Most building supply stores will even provide a blower unit free of charge when you purchase the insulation from them. However, many homeowners simply carry bags of material to the attic and simply pour and rake the material flat.

There is very little difficulty in adding insulation to the attic. The major concern is to ensure that attic vents are not plugged with the stuff. Take a look at Figure 3-30 and see how cardboard baffles have been stapled or pressed into the space between the rafter or joists. These baffles prevent insulation from rolling down into the soffit vent area.

Ensure that insulation is not applied around electrical light fixtures or bathroom vent fans that are not specifically approved for insulation coverage. Overheating light fixtures are a sure way to start a house fire.

Appliance Selection

What's all the Fuss About?

A home without major appliances, computers and entertainment systems is a life style that few people would embrace. There is nothing fundamentally wrong with having air conditioning or a large screen TV (other than watching 18 hours of it a day). The use of energy to power these devices is not in itself a problem. The problem occurs when homeowners choose to use polluting, non-renewable resources as their main energy source to power wasteful appliances. There is no free lunch. At some point in the future, (some would say right now) these non-renewable resources will start giving up (remember the blackout of 2003?) or kill us with their exhaust.

The good news is that governments and appliance manufacturers

Figure 3-31 A standard Sears 18.5 cubic foot, 2-door refrigerator like this one will save hundreds of dollars and kilowatt-hours of energy over its 20+ year operating life.

are beginning to understand these issues. Manufacturers are actively working to lower the energy requirements of their products. For renewable energy system users, these improvements have exponentially increased the number of appliances available to them. For energy conservers on the grid, these same appliances are helping put money in the bank and assisting you with living just a little bit lighter on the planet.

The average 25-year-old refrigerator (the one keeping a 6 pack of beer cold in the basement) uses approximately 2,200 kilowatt-hours of energy per year to operate. Electrical rates vary greatly around North America, however the average daily rate in California (when there is power) is approximately 15 cents per kilowatt-hour or $330 per year operating cost [1]. A new Sears 18.5 cubic foot, 2-door unit uses 435 kilowatt-hours per year or $65.25 to operate, at a savings of $265 per year. At that rate, it would only take 3 to 4 years to pay for itself (assuming rates don't climb any higher). Put another way, if you translated the savings into beer, you could probably supply the suds for the whole block on the savings alone! If that isn't incentive to switch, what is?

For those of you who want to produce your own electricity, the evidence is even more pervasive. Remember we talked about the U.S. Department of Energy? They state that for every $1.00 you spend on conservation, you will save between $3.00 to $5.00 in electrical generation equipment. Not only do you save energy, you reduce electrical equipment costs by up to five times!

1. Rates fluctuate from a low of 3.3 cents per kilowatt-hour in Ontario to peak daytime rates in California of 30 cents (all prices in US$). Check your utility rate (per-kilowatt-hour) and multiply it by the yearly consumption rating of your current and proposed appliance models to calculate the savings.

How to Select Energy Efficient Appliances

It's not necessary to carry around an energy meter, like the one shown in Figure 3-32, to measure the energy consumption of each appliance you wish to purchase, although it wouldn't hurt.

These meters (which are available from many sources, including www.theenergyalternative.com) plug into a wall outlet, and the appliance plugs into a receptacle on the front of the meter. The meter display indicates the electrical power consumed. Remember that power (watts or W for short) is a measure of the flow of electricity (amps or A) multiplied by the pressure of electricity (volts or V). Most major appliances such as washing machines, refrigerators, dishwashers and food processors plug into a standard 120 V outlet

Figure 3-32
An electronic energy meter such as this one will tell you how much power an appliance requires to operate. Most models can even be programmed with your utility rates to tell you cost, over a period of time.

(the pressure). The flow of electricity in amps results in the wattage. For example, a typical food processor draws 2.4 A and plugs into the wall outlet:

120 V x 2.4 A = 288 W

On the other hand an electric kettle draws 12.5 A:
120 V x 12.5 A = 1,500 W

So, what does this mean? Let's suppose that you and your partner are shopping for a new television set. You compare all of the models and find two that are about equal on your list of requirements. You whip out the power meter, or if you are more conservative, authoritatively inspect the electrical ratings label and find that model "A" uses 162 W of power while model "B" requires 1.9 A.

Now, at this point most people would run screaming out the door having flashbacks to those grade nine "A car is moving west..." type problems. You, on the other hand, have studied *Chapter 2 – Energy 101* and recall that power (in watts) is the voltage multiplied by the current.

120 V x 1.9 A = 228 W

You close the deal by smoothly informing the sales person that model "A" is the better TV as it will save you loads of dough over its operating life, reduce green house gas emissions and require less electrical generation equipment if you were using off-grid energy sources. You might even save money on the purchase, with the sales person trying to get rid of you.

But power is only half of the equation. If for some strange reason you were to watch model "A" television 2 hours per day and model "B" for 1-1/4 hours per day, the energy consumption for model "B" would then be lower:

Model "A" energy consumption = 162 W x 2 hours = 324 W-hours
Model "B" energy consumption = 228 W x 1.25 hours = 285 W-hours

Time factors into the energy cost. This is why we want to turn off lights in empty rooms.

There has to be a Better Way

Most people are not so fanatical as to carry an energy meter with them when they go shopping for appliances. Some people may take a look at the electrical ratings label, provided they can keep the watts and volts straight without having to carry this handbook.

For the rest of us, the government has made our life a little easier, at least for the larger appliances. Figure 3-33 shows a Canadian "EnerGuide" label that is affixed to a high-efficiency washing machine. The EPA uses a similar label and program called "EnergyGuide". Both programs require that labels be affixed to appliances and provide comparison data for appliance models of similar size with similar features. It also indicates the appliance's total energy consumption per year.

A closer look at the label reveals the following features:

- The bar graph running from left to right represents the energy consumption of all models of similar appliances

- At the left side of the graph is the energy consumption in kilowatt-hours per year of the most efficient appliance.

- At the right side of the graph is the energy consumption of the worst appliance in the same class.

Referring to the bar graph on the EnerGuide label, we can see that the most efficient appliances' energy consumption is 189 kWh per year and the least efficient is 1032 kWh per year. Think about that for a moment. Two appliances of the same size and

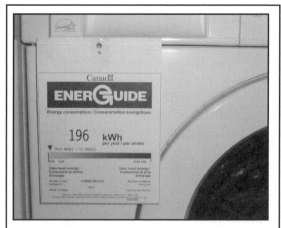

Figure 3-33 This high-efficiency washing machine carries both the "EnerGuide" and "Energy Star" program logos, allowing consumers to quickly determine energy consumption between models.

class, with one consuming over five-and-a-half times the amount of energy to do the same job! In dollars and just plain sense this equates to:

Efficient washing machine = 189 kWh per year x 10 cents per kWh = $19 per yr.
Inefficient washing machine = 1,032 kWh per year x 10 cent per kWh = $103 per yr.

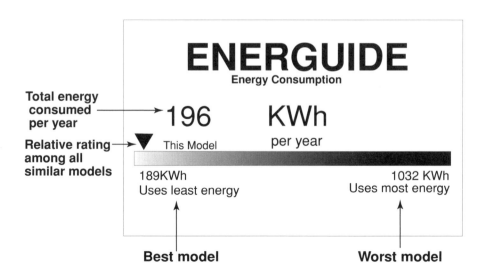

Figure 3-34 Detailed view of the EnerGuide program label.

Figure 3-35 This Sears high efficiency clothes washer uses five-and-a-half times less energy than a similar sized unit. Coupled with the gas dryer to the left, this makes an excellent choice for any off-grid, renewable energy system.

Assuming 10 cents per kilowatt-hour of electricity, the difference in operating expense is $84 per year. Over the life of the machine (say 20 years), that's a whopping $1,700! For those people taking the off-grid plunge, the difference just can't be ignored.

In Figure 3-34, the triangle-shaped pointer over the bar graph on the EnerGuide label shows this particular appliance model's energy consumption in relation to the most and least efficient models of the same class. The closer the pointer is to the left side of the bar graph, the lower the operating costs.

One interesting point to consider is the increase in cost of the appliance as you near the most efficient side of the graph. The sample label used in Figure 3-34 is on a Sears front-loading washing machine, straight from the catalog pages. The most efficient model is the Staber horizontal-axis machine. The Staber is almost twice as expensive as the Sears model but only one percent more efficient. On the other hand, the least efficient model is marginally more expensive than the Sears high efficiency model.

Small Appliances

The EnerGuide program was designed to take care of major "white goods" appliances. What about smaller appliances and electronics that don't carry the program label?

Our choices at this point become a little bit more difficult. We have to revert back to using an energy meter or reading the electrical ratings label of the device. An electrical meter such as the one shown in Figure 3-32 is a great device, provided you have access to the appliance you wish to purchase for a test analysis. The reason this matters has to do with energy consumption, which by definition requires time to evaluate (energy = power x time).

Let's take a look at an every day coffee maker to see how time plays into our assessment. A typical coffee maker such as the one shown in Figure 3-36 has an electrical rating label showing 120 V and 10.5 A. Whipping out your calculator you correctly arrive at the power consumption of 1,260 W.

Figure 3-36 A coffee maker uses a lot of energy to boil water, but considerably less to keep the pot warm. Use care when calculating energy consumption from the manufacturers label.

120 V x 10.5 A = 1,260 W of power

At 7:00 AM on Saturday morning, you stumble down the stairs and get the brew going. By 11:30 you have slugged back your third cup, draining the machine and shutting it off. Applying your caffeine honed mathematical skills, you determine that the coffee maker was on for 4 1/2 hours, giving you an energy calculation of 5,670W.

1,260 W of power x 4.5 hours = 5,670 watt-hours of energy or 5.7 kWh

At 10 cents per kilowatt-hour for energy, you quickly realize that your morning coffee just cost you 57 cents worth of energy, right?

5.7 kWh x $0.10 per kWh = $0.57

Wrong. The math is correct, but the assumptions are wrong. You have to be very careful when calculating the energy consumption of an appliance as it may in fact change with time. Yes, the coffee maker label did say that it required 10.5 A or 1,260 W of power. What it did not tell you was that it only needed that much power to boil the water. Once the coffee was brewed, it uses less power to keep the pot warm.

So what is the correct answer? Actually, I don't have a clue. Without access to the coffee pot and an energy meter to try it for a while, it's anybody's guess. When trying to calculate the energy consumption of any small appliance (i.e. those without an EnergyGuide label) follow these general guidelines:

- Appliances that draw a lot of wattage, typically over 300 W, should not be used in an off-grid home unless it is used only for a short period of time (i.e. less than 10 minutes) or during periods of excess energy production. This applies to crock pots, coffee makers, hair dryers, curling irons, electric kettles, clothes irons, car block heaters and space heaters. For those of you on the grid, keep the same rules in mind; you will save energy, money and the planet to boot.

- Coffee makers such as the model described above can be used, but transfer the fresh coffee to a thermos bottle server. The coffee stays warm and tastes better, without using unnecessary energy.

- Consider boiling water using a propane stove and using a drip coffee basket.

- Refer to *Consumer Guide to Home Energy Savings* for electrical consumption ratings of desired appliance models before you buy. (The book is available through Real Goods (www.realgoods.com), part number 82-399).

- Refer to Appendix 2 for a listing of various electrical appliances and tools. Typical power and energy calculations are provided for many products.

- All major heating appliances such as cook stoves, ovens, electric water heaters, electric clothes dryers, furnaces, central air conditioners, etc. cannot be economically used in an off-grid home due to the enormous amount of energy consumed. In most of these instances, switch these appliances to propane.

- Appliances that use 240 V connections are not generally suitable for an off-grid home. Most of these units are large energy consumers, which we don't want in the first place. The sole exception might be a 240 V submersible well pump, which we will discuss later.

- Microwave ovens are acceptable; provided they are used within reason (refer to the section below discussing phantom loading).

- For off-grid homes, try to purchase appliances that are not equipped with an electronic clock or "instant-on" anything. These devices are considered phantom loads and consume a large amount of power without doing anything for you. We will review what to do with these loads in *Chapter 5 – A Re-*

Figure 3-37 A satellite dish brings a welcome distraction during winter storms. Just remember to switch off the power input using a special plug or power bar to eliminate phantom loads, discussed in Chapter 5.

newable Electricity Primer.

- Convert as much lighting as possible to high efficiency compact florescent lamps (see below).

- Use high efficiency front-loading washing machines. The Staber HXW-2304, for example, is so efficient that you only require one ounce of soap per wash load. You will save electrical energy to operate the washer, less water per cycle, and use less heating energy to heat the smaller amount of wash water. How can you go wrong?

Figure 3-38 A wall full of home theater equipment may not seem very energy efficient, but in fact it is. Modern electronics are a wonder of efficiency and work well with any renewable energy system.

- If you are not sure about an appliance or tool, borrow one from a friend and plug it into an energy meter like the one shown in Figure 3-32. Even if the appliance model is not exactly the same, it will give you a general idea on power and energy consumption. As a last resort, talk to your renewable energy system dealer or system designer for advice. If your home is off-grid, it should have an energy meter built into the power system.

Computers and Home Office Equipment

Many people are finding that self-employment is the way of the future, especially after the last round of corporate downsizing. An essential part of working and playing at home now revolves around computers and related home office equipment.

With your computer, desk lamp, printer, modem and monitor working overtime (maybe all the time if you have teenagers), how will this impact electrical energy consumption?

Modern electronics are a marvel of efficiency and they keep getting better all the time. For example, let's compare a 7-year-old, 15-inch color computer monitor with a new flat screen model. The older unit may use as much as 120 W of power, while the new one draws about 30 W. That's a 400% reduction in power usage.

You should also consider that few people use a computer for only a minute or two, and then turn it off. The average user may have the computer and support system operating for many hours, which increases energy consumption (power x time). It makes sense to purchase the most efficient products because of this longer operating time.

Don't just think about the computer. You will almost certainly have a printer, monitor, and desk lamp, room lights, fax machine and possibly a photocopier to fully equip your home office. Teenagers (and aging rockers) can add to this list a 400 Watt surround sound system. Running an energy efficient system for 40 hours a week will still burn up a lot of juice.

Fortunately, the government has come to our rescue once more. The Energy Star program has been adopted across North America and is the equivalent to EnerGuide and Energy Guide labelling. The Energy Star program provides guidelines and re-quirements to manufacturers who seek compliance for their data processing products. Most compli-ant products utilize high efficiency electronic components and special-ized power saving software. For ex-ample, a compliant monitor will en-ter a low-power sleep mode if the image has not changed within a given timeframe. Other examples include laser printers adjusting their heater temperature when idle.

Money Isn't All You're Saving

Figure 3-39 Purchasing computer and home office equipment with the Energy Star label ensures the lowest energy usage.

As with any appliance, some products are better than others. A laptop compu-ter will always be more energy efficient than the equivalent desktop model, owing to the limited energy stored in the batteries. Likewise, a bubble jet printer is more efficient than a laser printer.

As with everything else in life, there are choices to be made. If you want lower per print costs, then laser printer toner is cheaper than bubble jet cartridges. Laptops are usually more expensive than a desktop system. Just remember to shop the en-ergy labels and look for the Energy Star logo.

Energy Efficient Lighting

Possibly the single most important invention to touch our lives is the incandescent light bulb. Prior to Mr. Edison's 1879 discovery, it was almost impossible to stay up past sundown (my wife still has this problem). Life before the modern light bulb meant filling a kerosene lamp and enduring poor lighting and air quality. In the early days of electrical power production, there was little thought to energy effi-ciency or environmental concerns. As a result, the bulb in your floor lamp has remained essentially unchanged for over 100 years. Times have changed and so have lighting technologies.

The incandescent lamp and its many variations create light by heating a small coil or filament of wire inside a glass bulb. Anyone who has watched a welder working with metal knows that hot metal glows a dull red color. If the metal is heated further, it glows brighter and with an increasingly whiter light. At some point, the metal will vaporize with a brilliant shower of white sparks. The various stages of glowing metal are known as incandescence.

In a similar manner, electrical power applied to a light bulb causes the metal filament to glow white-hot. If you decrease the amount of power reaching the bulb (by using a dimmer switch for example) the bulb dims and glows with a progressively redder color.

Figure 3-40 We owe a nod of thanks to Mr. Edison and an entire industry that developed a replacement for the kerosene lamp. As technology marches forward, it is now time to say goodbye to our old, but wasteful friend the incandescent lamp.

Making an incandescent lamp glow requires a large amount of energy to heat the filament. In a typical light bulb, 90% of the energy applied to the filament is wasted in the form of heat. Therefore, only 10% of the energy you paid for is making light, what a waste!

To make matters worse, we need to think about the heat component for a moment. If it is summer time, this waste heat contributes to warming the house and increasing the air conditioner load. An incandescent bulb hits you twice in the wallet: poor efficiency resulting in high operating cost and waste heat to make the light, as well as the added expense to get rid of the waste heat using air conditioning.

But does this waste heat and cost mean much? Let's put it into perspective. Assume it's a warm summer evening. You have a total of fifteen 100 Watt lights on in the house. The waste heat is:

15 bulbs x 100 W x 90% heat loss = 1,350 W of heat

This is about the same amount of heat output from a large electrical space heater used in the winter. That would be fine if it was winter, except that it is happening in the middle of summer!

But, you counter with: "This waste heat can be used in the wintertime to help warm the house." True, except that electricity is the most expensive means of heating a home anywhere in North America and this byproduct of lighting cannot be controlled unless you want to place your house lights on a thermostat. If you are generating your own electricity off-grid, this waste is not even manageable. So what are the alternatives and how do they save energy and money?

Compact Florescent Lamps

Mention florescent lamps to my wife and all she can think of are the pasty-faced girls applying makeup in the washroom at the high school prom. In the 60's and 70's some companies used the term "cool white" on their florescent lamps, as if this was a good thing.

Enter the compact florescent lamp (or CF lamp for short). These marvels of efficiency may look a little odd, yet they offer many advantages over standard incandescent lamps. To start, a CF lamp is designed to last ten times longer than an incandescent lamp. Shop around when looking for these lamps. Stores such as Wal-Mart carry the electronic CF lamps for about $3.50 each. Many stores still treat CF lamps as "special", including the pricing at about four times this cost.

Light output and energy efficiency is another consideration. Contrary to popular belief, light output is not measured in watts, but in *lumens*. The wattage rating of a bulb is the amount of electrical power required to make it operate. The standard 75 Watt light bulb gives off approximately 1,200 lumens of light. A 20 Watt CF lamp provides the same intensity, yet uses 4 times less energy. Translate this to cost savings and the CF bulb will save $55.00 over its life span, assuming energy costs of 10 cents per kilowatt-hour. This does not even take into account the savings in air conditioning load and environmental pollution.

Figure 3-41 The compact florescent (CF) lamp is the bulb of choice in modern, energy efficient homes. A standard 60 Watt incandescent lamp is shown at the left. An equivalently bright, 15 Watt CF lamp is shown immediately to the right.

Lastly, CF lamps won't give you headaches or the Bride of Frankenstein look first thing in the morning. Advances in phosphor coatings and electronic ballasting eliminate problems of ghastly color and flicker. The lighting industry uses a term called the Color Rendering Index (or CRI). All bulbs have this rating, although it can be a bit hard to locate the information. The typical incandescent lamp has a CRI rating of 90 to 95, compared to a CF lamp with 82. Compare this with a Frankenstein florescent bulb rating of 51. Still not convinced? Go buy a couple of CF bulbs and try

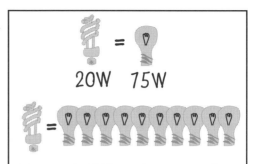

Figure 3-42 CF lamps last ten times longer than regular light bulbs and use approximately 4 times less energy. With color quality similar to incandescent light bulbs and no flicker, you won't get headaches or that awful pasty color when you look in the mirror.

them out. It will be almost impossible to tell the difference in color or operation once installed.

The Down Side

The shape and size of many CF lamps are not identical to standard bulbs. This may cause problems fitting the CF lamp in a conventional socket. Take note that shapes and sizes vary by manufacturer. Check several different brands to see if one will fit your application. Before resorting to changing the light fixture, see if your local hardware store is able to suggest an alternate base, harp or socket extension.

Dining rooms often have overhead fixtures that are connected to a light dimmer switch. CF lamps must never be placed in these sockets unless they are specifically designed for dimming. Be aware that dimmable CF lamps are available, but at a hefty premium cost.

CF lamps can be used in extreme outdoor winter lighting systems. In this application, the lamps typically require 1 to 2 minutes to warm up and produce useful levels of light. This is not a problem for security or perimeter lighting on timers. Garage door opener lights or those applications where the light is required quickly, should not use CF lamps.

T8 Florescent Lamps

Large area or kitchen cabinet indirect lighting often uses standard florescent lamps with magnetic ballasts. Although more efficient than incandescent lamps, these models are still only half as efficient as CF lamps or new "T8" florescent lamps with electronic ballasts. T8 lamps with electronic ballasts are more expensive than CF units, but are excellent replacements for older 4- and 8-foot standard florescent tubes. It's easy to recognize T8 lamps, as they are approximately half the diameter of conventional tubes. Likewise, electronic ballasts are very light for their size, especially when compared with older, less efficient magnetic ballasts.

12 Volt Applications

For those of you considering operating an off-grid cottage directly from 12 V batteries, T8 fixtures are now available for these applications. The Thin-Lite Company offers smaller fixtures that use electronic ballasts and run on 12 Vdc. There are a number of models to choose from with different light outputs and sizes.

There are also a number of 12 V LED lighting units available from www.realgoods.com. Although targeted for the marine industry, they also work in cottage and cabin applications.

A final word on 12 V lighting is to not make the same mistake as with 120 V incandescent lamps. The tiny "car taillight" bulbs or even flashlight bulbs you use are still incandescent. When you are considering 12 V lighting, energy efficiency is of the utmost importance.

Track and Specialized Lighting

There are places that even a die-hard energy conserver would never put a CF lamp. Lighting your Rembrandt or Picasso collection is one area that comes to mind. For these applications, low-voltage, high-intensity MR-16 lamps are ideal. Jewelry and watch stores use these miniature spotlights for clear intense lighting of small objects, artwork, and display cabinets.

MR-16 halogen gas incandescent lamps are fabricated with specialized internal reflectors that direct light in a unidirectional flood pattern. MR-16 bulbs operate on 12 V, which in turn requires a transformer (supplied with the light fixture) to drop the household 120 V supply. Their low wattage is similar to a larger CF lamp, so they can be considered efficient in these special applications.

Figure 3-43 MR-16 style flood lamps are excellent for lighting your priceless jewelry or baseball cap collection. Although they are incandescent lamps, their low-wattage and high intensity make them acceptable for these applications.

Large Area, Exterior Lighting

Large outdoor areas require major lighting muscle. While it is possible to use a few million CF lamps to light your yard, an easier, even more energy efficient way is to install high or low-pressure sodium lamps. Sodium lamps have a characteristic yellow color that makes them suitable for perimeter and security lighting. They also work well in horse riding arenas and for street lighting and so forth. Watt-for-watt, a sodium lamp produces twice as much light for the same power as a CF lamp (and they are up to ten times more efficient than incandescent lamps). For those of you operating off-grid, sodium lamps are an excellent choice for lighting the kids' hockey rink or the barnyard area.

Free Light

Perhaps the best lighting is free…100% energy efficient. No, don't just open the drapes; that would be obvious. Even the best-designed house may have an area or two that suffers from low light levels during the daytime. These areas may require lighting even on the sunniest days.

For these areas, a review of the skylight is in order. Typical skylights are not energy efficient, are prone to condensation problems and often require reframing and finishing from the inside. A product available from www.sunpipe.com offers an excellent alternative solution. The Sunpipe is similar to a fiber optic device on steroids. An intake piece is mounted on the roof in the same fashion as the metal chimney shown in Figure 3-44a. A reflective supply pipe is fitted down through the roof opening into the living area. A trim kit is installed and that's it.

Figure 3-44a. This view details the "intake" portion of the Sunpipe unit.

Figure 3-44b. The hallway is completely illuminated with a Sunpipe. Covering the intake to show the difference without the Sunpipe, created an almost black photograph.

Once installed, the Sunpipe directs outside light into the immediate area, even on cloudy days. An example installation is shown with the Sunpipe illuminating a narrow second floor hallway in Figure 3-44b. If the Sunpipe were to be covered, this hallway would be almost completely dark. A 9-inch diameter pipe provides the equivalent illumination as a 400 W incandescent lamp on a sunny day.

System Sizing

Let's finish up our electrical conservation section with a little homework by putting some of this theory into practice. Start by checking your major electrical loads in the house. Anything that plugs into a wall outlet has a label on it that tells you the voltage (either 120 V or 240 V) and current or wattage for that item. Although these labels often over-state the energy usage, they are useful as a guide for calculating energy consumption. Write down the wattage data from these labels into the *Energy Consumption Worksheet* in Appendix 7 (If the appliance label does not show wattage, multiply the amps x volts to calculate the watts). The next step is to see if you can estimate how many hours you use the device per day. If you only use a device occasionally, try to estimate how long it is used per week and divide this time by seven. A quick fly through the calculator and, *voilá!*, your "daily average energy usage per day" is computed. Keep this number handy. It will be required as we start to build our renewable energy electrical system starting in Chapter 5.

Water Supply and Conservation

Introduction

Turning on a tap for a glass of water is a luxury that few people give any thought to. If you live in the city, all it takes is writing a check to your local utility a few times a year and presto! Clean, clear, life-giving water sprays forth.

Society's lack of concern or ignorance for this natural resource is often beyond comprehension. One only has to drop by a golf course in the desert to see first hand the waste of a precious resource. Many people counter with: "Why not use water as we wish? Isn't water a renewable resource, like the solar energy we are capturing?"

Figure 3-45. An average person in North America consumes 71 gallons (270 liters) of water per day. The conserver can easily reduce this consumption by 50%.

Water is a plentiful and renewable resource, within limits. The majority of the earth's water is stored as moisture in the air, salty seawater and locked in ice and snow. The balance of the remaining fresh water can be had by drilling wells to reach the underground water table or from lakes and streams.

Lake and ground water levels have been dropping for many years, with hardly a day going by without some newspaper headline screaming about the subject. Part of the reduction in water levels is due to human population increase and usage. Industrial and climatic changes are also being blamed for an ever-increasing reduction in our second most important natural resource after air.

The Conserver Approach

An average person in North America consumes 71 gallons (270 liters) of water per day. Besides saving money on water bills and waste heating expenses, reducing this consumption will lighten the load (electrically and literally) on water and sewage treatment facilities. Water is also very heavy, requiring large amounts of energy to operate pumps in rural homes supplied by wells.

It is estimated that 75% of water usage is in the bathroom, with 20% being used for laundry and dishes. The remaining 5% is used in cooking and drinking. Conserving water and the energy used to heat and move it requires using the same approach as used with lighting: do more with less. This does not mean that the whole family has to get in the shower at the same time. What it

Figure 3-46. This pie chart shows how the average person consumes water.

does mean is using more efficient appliances and methods to achieve the same results.

Harvesting our Water Supply

The best place to start is in collecting our water supply. For those of you connected to municipal water supplies, you might want to jump ahead to the next section, *How to Conserve.*

The two most common methods of collecting water are pumping from a well and pumping from a surface supply such as a lake or cistern. Let's start with the well, as this is the most common method in use.

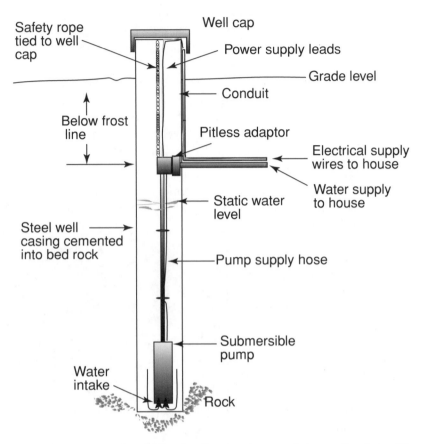

Figure 3-47. This diagram details the installation of a typical deep well submersible pump.

Deep Well Submersible Pumping

Referring to Figure 3-47, you will see an installation diagram of a typical deep well submersible pump. For any new or retrofit application, the deep well submersible pump is the simplest, most reliable and easiest-to-service design. The deep well submersible pump is available as a standard 120/240 Vac centrifugal unit or in high-efficiency designs such as the Dankoff ETA series pump shown in Figure 3-48 (www.dankoffsolar.com). Both designs install in the same manner, the major difference being the technology used in the pump design and resulting efficiency. The Dankoff submersible pump is so efficient, that it can be used as a remote well pump for livestock with only one small photovoltaic panel powering the unit as shown in Figure 3-49.

Figure 3-48. The Dankoff model ETA090 submersible deep well pump installs using the same techniques as regular AC powered well pumps. (Dankoff Solar)

Figure 3-49. The Dankoff model ETA090 can be used for standalone pumping applications such as this remote livestock watering well. Very efficient, but not nearly as picturesque as the old western windmill. (Dankoff Solar)

For off-grid applications, efficiency is important; however, many homeowners (including myself) use standard AC submersible pumps. Although the efficiency is lower and the power requirements higher, they are less expensive and easier to find service parts in the middle of the night. Couple this with low water usage appliances and large storage/accumulator tanks and the resulting pump usage can be reduced to 10 minutes per day.

Slow-Pumping Techniques

The deep well submersible pump is the brute force method of pumping water for a home pressure system. The "do more with less" method involves a different approach that is not nearly as popular as deep well pumping, but works just as well.

A slow-pump is simply an ultra high-efficiency, low-voltage pump that has low-flow characteristics as compared to a standard deep well submersible pump. This lower flow characteristic requires a means of buffering water consumption by the addition of a storage tank. Figure 3-50 shows a typical slow-pumping submersible pump supplying water to a storage tank. A float located in the storage tank turns the submersible pump off when the tank is full and back on when the level drops.

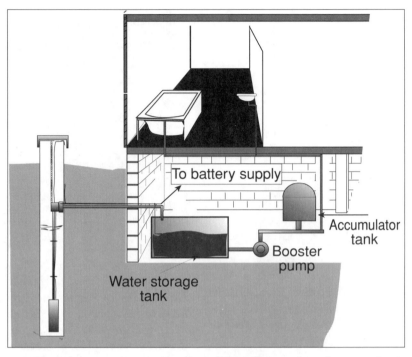

Figure 3-50. A slow-pumping system relies on high-efficiency, low–flow pumping into a storage tank to increase efficiency. This design reduces water pumping energy requirements by 50%. (Dankoff Solar)

A second pressure booster pump is installed between the water storage tank and the household supply. This pump pressurizes an accumulator (see below) and turns off when the water pressure reaches 40 to 50 pounds per square inch (psi). This is the normal maximum pressure for domestic water systems. The switch is also designed to turn the pump back on when the pressure drops below the 20 psi cut-in pressure.

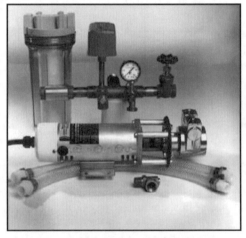

Figure 3-51. A booster pump can be used to pressurize water from a storage tank or from a lake or stream, provided freezing is not an issue.

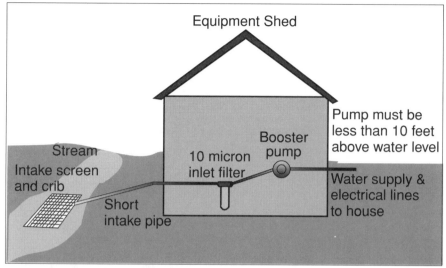

Figure 3-52. A typical surface water pressure system will operate in areas where freezing is not a concern.

Surface Water Systems

Surface water systems do not require a submersible pump unless the installation is subject to freezing. For temperate or seasonal use, the booster pump may supply the water and pressurization at the same time. Placement of the pump becomes of primary importance as pumps do not "suck" water, they push it. A typical installation is shown in Figure 3-52. An intake pipe is provided with a screened inlet (to keep the fish out) and supported inside a simple crib of rock or wood. The intake is then routed to a 10 micron sediment filter (to stop grit and sand, but not bacteria) from entering the pump.

The pump inlet must be positioned no higher than 10 feet above the water line to ensure proper operation. The booster pump then supplies pressurized water to the accumulator (see below). A built-in pressure switch turns the pump on and off at the desired operating pressures in the same manner as the slow-pumping method described above.

Installations such as the one outlined in Figure 3-52 can vary greatly. One area to be aware of is the vertical height or *head* distance that water must be pushed up hill. Assuming that your cottage or home is at approximately the same level as the pump, there is no problem. For houses that are located above the pump, water pressure cut-in and cut-out settings must be increased by 1 psi for every 29 inches (74 cm) in vertical lift, assuming the pressure switch is located at the pump.

This change in pressure switch setting accounts for the weight of water in the pipe and its affect on water pressure above it. For example, if the house is located 13 feet above the level of the pump, the pressure switch settings will have to change by:

(13 feet above pump x 12 inches per foot) ÷ 29 inches per psi   5.5 psi incr ease

If the pressure switch is located in the house, this change is not required although you will have to be sure the pump can work with the increased pressure requirement. In addition, be aware of the maximum vertical height a pump can push water and still supply adequate water pressure. It's fine for the manufacturer to say that the pump delivers 50 psi, but if the cottage is 120 feet above the lake surface, the pressure at the top won't be enough to water your pet gold fish.

(120 feet above pump x 12 inches per foot) ÷ 29 inches per psi   49.5 psi used up

Pressure Accumulation

Anyone who has been to a rural home or cottage with an improperly installed water pressure system knows the rub. The shower pressure varies from a trickle to you being blown against the wall. Getting a drink of water means sometimes waiting two minutes to fill the glass, while the next time, the whole kitchen gets flooded from the spray. During this time, the well pump is kicking on and off like a wild Mustang being broken in.

An improvement over this comical water supply system is to even out the flow and pressure by installing a proper water accumulator tank. When your plumber offers you a coke bottle sized accumulator, take the refund and purchase one that works. The accumulator shown in Figure 3-54 is enormous compared to the laughable units normally supplied. This model—the water accumulator,

Figure 3-53. An intake sediment filter is required for all slow- and booster-pump applications. This filter is ready to change as sediment has accumulated in the filter housing.

not Lorraine—holds approximately 40 gallons (152 liters) of water and provides an even flow throughout the longest shower (teenagers notwithstanding).

In addition to smoothing out the water supply, the large tank greatly reduces the pump cycling by storing most of your daily water usage. After applying all of the water conservation rules outlined below, Lorraine and I often manage with the pump only cycling once per day.   A further advantage of the large accumulator is water temperature buffering. Well water is very cold, often entering the house at 48 degrees (9 Celsius). This water has to be heated, often with propane. The accumulator acts as a large heat sink, absorbing room air heat (remember from Chapter 2 – *Energy 101* that heat moves from warm to colder areas?). This free heating will warm the water to near room temperature, saving more money by lowering the amount of fuel needed in the water heater.

Water Supply Summary

There are dozens of methods available for installing a domestic water supply system, many of which require specialized design support. If simplicity and ease of service are primary concerns then stick to standard deep well submersible AC pumps with large accumulator tanks. This is pretty standard stuff that any plumber can understand and, when used with proper water conservation, it is efficient

Figure 3-54. A large water accumulator provides city-grade water pressure and flow.

enough for all but the smallest of renewable energy, off-grid systems.

How to Conserve

Now that we have a supply of water, let's see how to make the most of it. Referring back to our pie chart in Figure 3-46 will show us what activities are the highest consumers of domestic water are.

Toilet Flushing

Tackling the task of conserving water when toilet flushing is not only simple, it is actually the law in many areas wracked with inadequate water supplies. Simply remove the old clunker and replace it with a certified low-flush model. This conversion will reduce water usage by 66% or more. Conventional toilets can consume over

Figure 3-55. The low-flush toilet requires only 33% of the water required by earlier models.

4.7 gallons (18 liters) per flush while low-flush models use only 1.6 gallons (6 liters).

Low-flush toilet models are available in a mind-numbing array of styles, colors and shapes to suit even the most discriminating derriere. Costs keep coming down, with economy models available in the $50 range. Another good reason to consider these models is the lack of condensation dripping from the tank. If your old clunker drips every time the humidity rises, that's the excuse you need to make the swap.

No matter which brand of toilet you own, if it leaks, it will be a water waster. A toilet that leaks 4 gallons of water per hour (16 liters) would fill a large swimming pool if allowed to continue for one year. Not sure if the toilet leaks? Place a couple of drops of food coloring in the water reservoir. Check the bowl after 15 to 20 minutes and see if the color shows up. If it does, the next book you need is "An Introduction to Plumbing" to fix it.

Bath Tubs

If you and your family are the types that prefer bathing, there is not much you can do except change an older bath to a new fiberglass model. Old cast iron or metal baths that are 6 feet long and very deep require a lot of water and heating fuel to fill them up and keep them warm.

A fiberglass soaker bath can be ordered in a 5-foot model that is fine as long as you aren't related to Magic Johnson. A shorter bath requires less water volume and the fiberglass model resists heat loss through the sides.

If you are having a bath during the heating season, leave the water in the tub until it cools down, then pull the plug. This little trick will transfer the heat energy from the water into room, providing you free warmth and humidity to boot.

Showers

Aside from turning off the water pump when the teenagers decide to live in the shower, the best solution is to add a low-flow showerhead. These units, such as the one shown in Figure 3-56, have built-in water flow restrictions and special "needle orifices" to reduce flow while at the same time making it feel like you just aren't standing under a leaky roof. Look for models that carry a certification agency displaying a flow rate of less than 2.6 gallons per minute (10 liters per minute).

Figure 3-56. Showerheads such as this model reduce water consumption by 50% and, don't forget, savings in hot water heating costs.

The Kitchen

A very quick update to the kitchen sink may be in order. Try replacing a standard flow faucet for one with an "aerator" nozzle, such as the one shown in Figure 3-45. These devices mix air with the water stream, giving the appearance and feel of more water flow. This apparent increase in water flow offers more coverage when washing dishes and hands.

Dish Washing Machines

Like every other major appliance in the house, dish washing machines are subject to the government's EnerGuide and Energy Guide programs. When purchasing a new model or updating a tired machine, review the EnerGuide or EnergyGuide label and purchase the most efficient model available. Better yet, try washing your dishes by hand. The savings in energy with "manual" models is quite high, and you might find this a good time to catch up on the family gossip.

Clothes Washer

The clothes washing machine is in the same boat as the dishwasher. Check the EnerGuide or EnergyGuide labels and purchase the most efficient machine you can afford. The washing machine shown in Figure 3-35 consumes five-and-a-half times less energy than a similar new machine. Shop carefully!

Water Heaters

Besides changing your water heater completely, there are only a few things that can be done to help conserve energy.

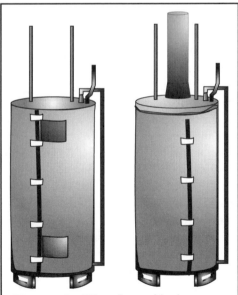

- Turn down the thermostat of the water heater to the lowest temperature you can accept. The thermostat on most gas units is visible and well marked. Electric units often have two thermostats underneath a removable cover. When adjusting them, ensure that the power to the unit is turned off or if you are unsure, contact your plumber for assistance.

- Install an insulation blanket as shown in Figure 3-57. These blankets are available from most building supply stores or plumbing contractors. The insulation batting helps to stop

Figure 3-57. Water heater blankets save a great deal of energy lost in hot water storage tanks. Make sure the blanket does not block any air intakes or vents on gas water heaters.

heat transfer from the hot water inside the unit. If you create a homemade blanket, make sure it is fireproof and that it does not block any air intakes or vents on a gas water heater.

• Install pipe insulation on all hot water lines. This material is very easy to install (where the pipes are exposed) and is inexpensive to purchase.

• Add an active solar heating system to capture the sun's rays to warm the water in your tank.

Water heaters have a lifespan average of ten years. Before you discover a newly installed indoor swimming pool courtesy of your leaking unit, perhaps its time to consider an upgraded model. Besides, units that are reaching their useful end of life are a bit like refrigerators. The savings in energy efficiency in newer models will easily offset the installation cost.

Figure 3-58. Instant water heaters such as this model in the foreground from Bosch do not waste energy by storing hot water all day long. They work by rapidly heating the incoming cold water, supplying enough hot water for the entire house. Note the standard water heater in the background, used as a diversion load for off-grid systems.

New to North America are the instantaneous water heaters such as the Bosch model shown in Figure 3-58. These units have long been used in Europe where energy costs prohibit wastefulness. For off-grid systems, these units do not require any electricity to operate either. Your plumber can easily replace your old storage unit and give you back a bit of floor space at the same time.

A standard propane water heater that is very effective in off-grid systems is the State (www.stateltd.com) storage unit, shown in Figure 3-59. Although it is no more efficient than equivalent new models, it does offer the advantage of being directly vented and does not require any electrical power as is typical of similar models.

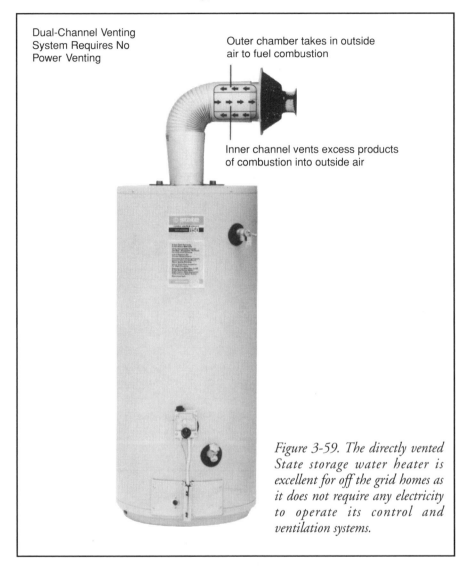

Dual-Channel Venting System Requires No Power Venting

Outer chamber takes in outside air to fuel combustion

Inner channel vents excess products of combustion into outside air

Figure 3-59. The directly vented State storage water heater is excellent for off the grid homes as it does not require any electricity to operate its control and ventilation systems.

Solar Hot Water Heating

Heating water is one of the most efficient uses for renewable solar energy. Energy capture and conversion rates may reach as high as 70% for water heating usage. Contrast this with the best photovoltaic panels (which convert sunlight to electricity) at approximately 15%.

A solar water heating system is comprised of a number of elements which capture and transfer heat energy to a storage tank. There are numerous means of interconnecting the various components, but a typical system is shown in Figure 3-60.

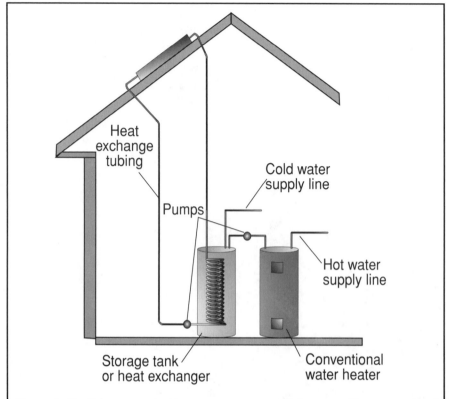

Figure 3-60. Solar water heating systems are one of the most efficient means of capturing energy from the sun. Perform a full evaluation with your heating contractor to ensure this technology is right for your application.

Direct systems, such as those available from Thermo Technologies Inc. (www.thermomax.com), involve cold water flowing into a solar collector plate and directly into the home hot water supply system.

Indirect systems utilize two distinct components as detailed in Figure 3-60. Hot water or an antifreeze mix is pumped through the solar collector to a heat exchanger and back again. The heat exchanger is in contact with potable domestic water, causing heat to flow from the solar collector system into the domestic supply. When the solar collector is cooler than the hot water supply, the distribution pump stops, preventing reverse heat flow.

Before contemplating such a system, contact a heating consultant or dealer to determine the feasibility of such a system for your application by performing a thorough evaluation of the economic payback.

Landscaping

Lawns

Possibly the most wasteful use of water (and energy) is used to maintain the perfect lawn. If there was ever a sinkhole for energy, this is it. We start by trucking in tons of topsoil, adding fertilizers, pesticides and lime to get it growing. Then we apply more fuel to cut, trim and edge our lawns. Added on top of this array of fun and games, is watering. Why water in the middle of a hot summer day and have 50% of the water evaporate or run along the sidewalk? If we must have perfect lawns, then we should try to water them more effectively.

Watering at dusk or early in the morning is best. Apply only as much water as your local conditions require. Check with a nursery or garden center and find out how much water is appropriate. Over spraying the lawn and watering the sidewalk is just plain wasteful. Spend the $20 on a better sprinkler unit and maybe a few dollars more on a timer. Program it to suit to the size and area of your lawn. If you are planting a new lawn, consider low-maintenance ground covers or drought-resistant grasses instead.

Are you in the market for a lawnmower? Consider a rechargeable electric model. Virtually no noise, minimal maintenance compared to gas models and no pollution. These units work very well as diversion loads in off-grid homes, absorbing those extra summer rays. What a perfect statement of working in harmony with nature!

Figure 3-61. Capturing rainwater is a time-honored tradition. Automate the process by connecting the barrels to a weeping hose for an automatic and labor-saving system.

Flower Gardens

The roof of your house captures a lot of rain during a downpour. Try containing some of this water in a series of rain barrels that are interconnected to supply flowerbeds. A few 40 gallon (200 liter) barrels interconnected with poly-pipe as shown in Figure 3-61 makes a nostalgic looking storage system. Connect the rain barrels together with weeping hose, just under the mulch of these beds, and you have an energy-efficient and labor-saving watering can. Remember that mulch doesn't just look good and suppress weeds; it also keeps the soil from drying out, further conserving water and energy.

Figure 3-62. Solar power garden lights won't help with water conservation, but they sure do accentuate the beauty of the flowers.

Although garden lighting does not help with water conservation, it helps to showcase your pretty flowers and hard work. This lighting will better harmonize with nature if it is solar powered. Solar-powered light prices have been dropping while their quality has steadily increased. The model shown in Figure 3-62 was purchased for under $5.00, and will continue to run long after you should be in bed.

Water Conservation Summary

The tasks described above will easily reduce your water consumption requirements by more than 50%. Add to this the reduction in electrical energy required to pump the water and the reduced water heating costs, soap consumption and off-grid energy system complexity, and you simply cannot justify not doing these upgrades. For your sake and the sake of the planet.

Sun-spiration

Leslirae and her daughter survived the Great Blackout of 2003…

"My daughter was scared and in the dark, as she couldn't find the flashlight and matches until I arrived home — after braving the traffic sans traffic lights, Thursday evening. I had anticipated such an occurrence, and had made sure I put fresh batteries in a flashlight and knew where I had left it, as well as having one book of matches with a striker panel on hand, and 3 pillar candles. I had more candles but had packed them away just a few weeks ago, thinking I would not be needing them before moving into a new house!

"We were lucky — although we had almost two days of "load-shedding" as it is called in India, the power came back in the middle of the night, before conking out for a shorter period on the second day. Fortunately we didn't lose much food in the fridge/freezer. The fridge was the only thing I'd left plugged in, and only because I could not move it to unplug it, to protect it from surges when the power returned.

The only fun part was trying to meet my deadlines for a client who was unaffected by the black-out (and who had asked me only a hour before the blackout if there was any reason I wouldn't be able to send her the document overnight) — I'd said "No, you'll have it in the morning'.

My laptop backup power died before I could finish writing the document by candle light (wax all over my desk!) and fax the document (e-mail was down), but as power was on in the morning, I got it done on time after all, albeit via early morning fax rather than overnight electronic format — so she had to retype to do the revisions!

I expect these problems will just get worse over time. We have been anticipating this for decades now but I was hoping to get the (renewable energy powered) house built before the blackouts started".

Leslirae Rotor

4.
Heating and Cooling with Renewable Energy

Heating and cooling are the largest energy consumers a homeowner is faced with. Everyone has a better place to put their dollars than in Big Oil's pockets. Chopping a few hundred dollars off your annual energy bill might just mean the difference between having to continue driving the old gasguzzling clunker and being able to trade it up for a zippy new energy-efficient hybrid car. Keeping the heat and cool air inside the house is the first step in your quest for energy efficiency (as discussed in Chapter 3 *Energy Conservation*).

Burning any fuel has some negative impact on the environment. Fossil fuels are non-renewable and contribute numerous chemicals and greenhouse gases—mainly carbon dioxide—

Figure 4-1. Trees are nature's solar collectors, using photosynthesis to convert atmospheric carbon dioxide, water, sunlight and soil nutrients into carbon, which is stored within the structure of the wood.

to the atmosphere. Some fuels are better than others. Natural gas and propane are the best, and low-grade coal is the worst. Consider that a natural gas or propane cook stove allows the burning fuel to vent *into* the house. Try *that* with oil!

Wood and wood pellets are different from fossil fuels in that they are renewable. A tree growing in the forest absorbs nutrients and water from the ground as well as carbon dioxide from the atmosphere. Photosynthetic processes within the tree convert carbon dioxide gas into carbon, which is stored in the structure of the wood. When we burn the dried wood, the carbon burns, releasing heat and carbon dioxide gas. The good news is that burning wood properly, exhausts no more greenhouse gas into the atmosphere than simply letting the dead tree rot on the forest floor. However, allowing even the best wood stove to burn smoldering fires adds pollution to the atmosphere.

If you are designing a new home, an even better energy source to consider is the sun. A properly designed and oriented home absorbs free energy on sunny days, helping reduce your reliance on burning anything. Both passive and active solar heating systems can be utilized. Passive solar heating is just that: passive use of properly oriented walls, windows and architectural house features. Active solar heating involves a complex series of solar collection and storage units, designed to increase the "density" of energy captured. Although active systems are a viable means of capturing solar energy, they are very complex and require far too many site specific details to be fully reviewed in this handbook. The resource guide in *Appendix 3* provides cross-reference information to other books and websites related to these systems.

The *Renewable Energy Handbook* covers the installation and operation of the following renewable energy heating sources:
- Passive solar systems integrated into the building design
- Heating with wood
- Heating with wood pellets

The pros and cons of different furnace and boiler technologies are discussed, as well as heat distribution systems that are compatible with renewable energy sources, including off-grid electrical systems:
- advanced combustion wood stoves using radiant heat circulation
- catalytic wood stoves using radiant heat circulation
- wood pellet stoves using radiant heat circulation
- hot water boilers using hydronic heat distribution
- backup heating systems

Passive Solar Heating

Solar energy is free, non-polluting and renewable. Why don't more people use it? My guess is that most folks think that houses using solar energy have to look like something George Jetson would live in. Or worse yet, a house designed by someone named Bucky Fuller. Well don't worry, geodesic dome designs aside, you won't have to upset your local town hall planning committee or be barred from block parties for life in order to live with a passive solar system. Another possible reason for being afraid of solar energy is its variability and the building contractor not being able to quantify it. If you require heating, just install a 150,000 BTU furnace then forget about it. Right? Everyone knows it's not quite that easy with solar energy.

Not so fast. Living with solar energy does not require a house full of complicated electronics and miles of glass. What passive solar energy does require is following some fairly simple guidelines to allow your home to take advantage of the sun.

Step One – House Orientation

Orienting the house to collect solar energy may seem obvious, but it's not. Most people give no thought to ensuring that the house is oriented in such a way as to capture as much sun as possible in the winter and provide proper shading in summer. This is how it's done:

- Using a compass at the proposed building site, locate magnetic south. Consult the magnetic declination chart in *Appendix 4* and determine the compass correction for true or solar north and south. Using this heading, place the long axis of the home at a right angle to it.
- Place deciduous trees in a path between the summer sun and the house. This provides shading for the house and glazing. When the autumn winds blow, the leaves fall from the trees, allowing the welcome winter sun indoors.
- Plant pine, cedar or other evergreen trees to the north and east to provide a wind block. In treeless areas, wind and storm blocks can be made using rock outcroppings, hills or even your neighbor's house.

These orientation strategies apply whether you are in the country or city. If you

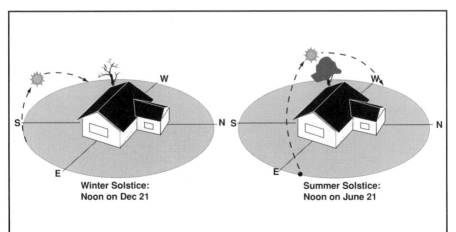

Winter Solstice:
Noon on Dec 21

Summer Solstice:
Noon on June 21

Figure 4-2. Orient the house to take advantage of winter solar gain and summer shading.

are designing a home in a city, try to evaluate the sun's path over your lot and take into account any shading effects caused by future buildings.

Step Two – Insulate, Insulate, Insulate

If you haven't taken the time to review *Chapter 3 – Energy Conservation*, go back and read it. A house that is under-insulated and prone to air leakage will quickly put an end to your quest for energy efficiency. **Remember that it is far less expensive to conserve energy than to purchase more energy to heat a substandard house.**

If you live in an area where summer air conditioning loads are high, consider adding radiant barrier insulation in the attic, as discussed in *Chapter 3*.

Step Three – Window Design

Windows are the passive solar collectors of the house. Vertically oriented glass will catch the low winter sun and miss the high summer sun angle. This means you can create as "normal" a looking home as you like and still get maximum solar efficiency.

- Place the majority of the desired glazing on the south side (long axis) of the house.
- Choose windows that are of the highest quality you can afford. Refer to *Chapter 3* regarding window construction. Remember that triple-glazed glass with Krypton gas fill are the most efficient available.

Figure 4-3. This home, located in Eastern Ontario, is designed to capture the maximum amount of winter sun energy.

Figure 4-4. This lovely veranda also forms the south window shading system in summer. The roof overhang was designed to allow sunlight to enter the house structure from mid-September through mid-March.

- Do not use low-E glass on the south side of the house or on any window that you wish to use as a solar collector.
- Install window units that can be opened on the side of the house facing the prevailing winds as well as on the side opposite of the house. You can open these windows to provide nighttime cross-flow cooling of the home.

Figure 4-5. This newly built "earth ship" design is very heavily glazed by northeastern standards. It will require plenty of shading and cross-flow ventilation to keep it from overheating in summer. (Note the photovoltaic panels mounted to the right of the windows.)

- Limit the amount of glazing on the north, northeast and northwest sides of the house.
- Do not over-glaze the south side. Excess window area equates to excess heat.

Step Four – The Finishing Details

These finishing touches will assist in making the job work "just right".

- Provide proper home ventilation using an air-to-air heat exchanger unit or alternative designs as outlined in *Chapter 3*.
- Provide sun shading over the south facing windows as required, to prevent the summer sun from heating the structure.
- Use summer sunscreen blinds similar to the units shown in Figure 3-7.

Heating with Electricity

The question "Why not use electricity to heat the house?" is based on the assumption that you can simply plug into your renewable energy system producers, such as wind and photovoltaic equipment, to make electricity.

Actually, you can heat a house with your renewable energy system, but the amount of electrical energy needed to do the task would require a wind turbine or photovoltaic system of enormous proportions. The installed cost of the renewable energy system would almost never pay for itself. Electric heating is also the most expensive heat source, even when you receive your power from the nuke down the street. For these reasons, renewable energy systems necessitate alternate energy sources to support large heating loads. For home heating, this means biodiesel fuel, wood and wood by-products (in the form of pellets). For large heat-producing appliances, such as cook stoves, water heaters and clothes dryers, the fuel of choice is natural gas or propane.

Heating with Renewable Fuels

Heating with wood goes back thousands of years. Europeans who settled in the New World along the east coast and in Pennsylvania experienced winters in North America that had a ferociousness never before seen in Europe. Rapid population growth coupled with poor energy efficiency in both wood-burning appliances and homes caused heavy deforestation and left towns with a grimy, sooty pall. It's no wonder that the majority of the population eventually picked up on the oil, natural gas and electric heat bandwagon. Write a check, set the thermostat and you have instant comfortable central heating....what could be better? As much as I would love to rant about the environmental impact of the oil-based house of cards, this is not the forum. So, sticking to the script, lets see why anyone would take a step "backwards" into the wood-heating scene.

Since about 10,000 B.C., there have been several advances in wood-heating technology. These advances have included new wood-burning stove designs that increase the amount of heat recovered and at the same time lower environmental pollution. Besides saving the environment, these technologies can also save your back: when you couple the energy conservation techniques discussed in *Chapter 3* with clean, high-efficiency wood burning appliances, you will reduce the amount of wood you have to cut, split, pile and consume.

When you have decided on wood as your main or secondary source of home heating, check with your stove supplier or local building inspector for installation details. Since the installation of wood and pellet stove systems are complex and vary from region to region, no specific installation details will be provided in this *Handbook*. It is highly recommended that your contactor or other professional installer do the installation work. Use the details provided in this section of the Handbook to increase your

Figure 4-6. This Vermont Castings (www.vermontcastings.com) catalytic wood stove easily heats this 3,300 square foot, ultra-energy efficient home. Located in the very cold reaches of eastern Ontario, it requires only two cords of firewood per heating season.

knowledge and understanding of wood burning to ensure you purchase the best stove to meet your needs.

Wood Burning Options

Burning wood is not just about tossing a log into a fireplace or refurbished wood stove from the antique dealer and expecting to heat your house. It takes a high-tech wood-burning stove to heat efficiently and for a long period of time. There are several types of stoves that fall into this category:

- advanced combustion
- catalytic
- biomass, wood pellet
- Russian or masonry heaters
- wood furnaces and boilers

The major concern of designers and manufacturers of these stoves is to safely and cleanly extract every BTU of available energy from the wood you load into the stove (remember BTUs and calories from Chapter 3?). The reduction in wood loading lowers costs by reducing the amount of wood purchased or cut and processed. The increase in efficiency comes by first creating an airtight burning chamber and then burning the smoke emitted by the fire before it reaches your chimney. (The masonry stove differs from this concept and will be discussed later in this chapter.)

An airtight burning chamber is required to limit the amount of oxygen reaching the fire. This lengthens the burning time and evens out the heat flow from the wood stove. Unfortunately, a slow, smoky fire also increases airborne pollutants and the risk of chimney fires, due to unburned fuel coating the chimney as creosote.

The idea of burning the smoke before it reaches the chimney may sound like I have been talking to a snake oil salesman, except it is true. Modern wood-burning appliance technology is aimed at developing ways of capturing the unburned fuel present in the emission of smoke. Smoke is the result of burning wood decomposing into "clouds" of combustible gases and tar. By applying additional oxygen and heat or special catalyzing materials, it is possible to burn it. The result is increased heat output and lower atmospheric emissions as well as a saving in wood fuel dollars.

Advanced Combustion Units

Advanced combustion wood stoves expand on the concept of a simple box stove that has been in use for over 100 years. Simple stoves allow the heat and smoke of the resulting fire to travel in a direct path up the chimney. The advanced combustion stove places an "air injection" tube into the smoke path. Secondary inlet air is drawn in through the tube increasing the oxygen content of the smoke, causing it to burn. The smoke then travels through a labyrinth path, radiating heat, before exiting via the chimney. This approach is like adding a turbocharger to a car's engine, free horsepower from normally wasted exhaust.

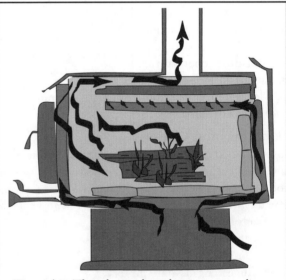

Catalytic Units

Once you understand how a catalytic wood stove operates, you can impress your friends by telling them how an automotive catalytic converter works.

Figure 4-7. The advanced combustion stove relies on secondary oxygen intake and smoke burning to increase efficiency.

The process really is the same except that in the case of a car, the unburned fuel is in the car's exhaust. Unburned wood smoke (or exhaust gases from a car engine) is routed through a catalytic device. The catalytic device is a ceramic disk fabricated with a myriad of honeycomb holes running through it. The disk is coated with a special blend of rare earth metals that have the unique feature of lowering the combustion temperature of the exhaust gasses, when mixed with a secondary supply of atmospheric oxygen.

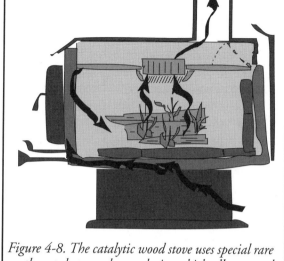

Figure 4-8. The catalytic wood stove uses special rare earth metals to produce a device which allows wood smoke to burn in the presence of a secondary air supply.

As the smoke (or exhaust) passes through the catalytic device, it is mixed with oxygen from an air inlet and ignites. If you watch the wood-burning stove's catalytic converter operating, you will see it glowing red hot and engulfed in flickering flames as the smoke is being consumed. A quick glance at the chimney on a crisp, cold January afternoon proves the operation with nothing but wisps of steam being emitted.

Biomass, Wood Pellet Stoves

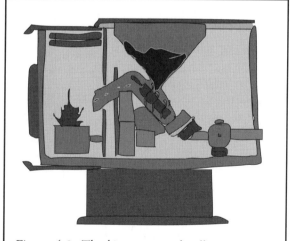

Figure 4-9. The biomass, wood pellet stove uses a motor driven auger to feed precise amounts of waste wood pellets into a fire pot.

Pellet stoves utilize biomass and waste wood by-products from manufacturers of furniture, lumber and other wood products. The waste material is ground and pressed together using naturally occurring resins and binders to hold the (rabbit-food like) pellets together. As a waste product, biomass fuel pellets offer excellent synergy, heating your home while reducing landfill waste at the same time. Pellets are convenient as they are supplied in neat and compact dog food sized bags, which can be stacked in your garage, ready for use. Simply scoop a bunch into the hopper of the stove about once per day and the controlled feeding unit will automatically deliver the right amount of fuel to the burner. The only down side is no electricity, no fire. For off-grid systems, this in itself does not present a problem; however, electrical loading of the small motor and control system must be considered, as discussed in *Chapter 5 – A Renewable Electricity Primer Course.*

Russian or Masonry Units

When we think of Russia, we think winters in Siberia. In that neck of the world, serious heat is needed, and there is no fooling around with wimpy stoves. The masonry heater is a serious unit. It is designed such that the firebox and chimney are lined with refractory brick. The firebox flue is routed through a labyrinth path, designed to slow the smoke on its way to the chimney. An external facing of brick, stone or adobe completes the design and increases the mass of the unit.

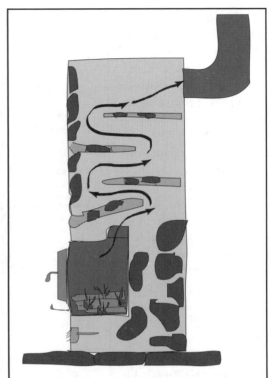

Figure 4-10A/B. A Masonry stove's size makes it the centerpiece of the house. A popular design from the Old World, they provide heat long after the fire has burned down.(Temp-Cast Inc.)

When these units are operated, a fast, furious fire is ignited in the firebox. Smoke and its accompanying heat zigzag through the flue passages, giving up heat to the surrounding masonry work. Depending on the locale, one fast fire per day is all that is required to heat the masonry . After the fire burns down, the stonework radiates its stored heat into the house.

The masonry unit achieves its environmental passing grade by creating a hot, fast fire. Fast firing of a stove will generate the same amount of heat energy as a slow fire, but in a shorter amount of time. The resulting hot fire consumes more oxygen than a slow smoldering fire and burns more completely, resulting in fewer pollutants.

The down side of masonry heaters is their cost, the need to keep re-firing and the units' size. Assuming you have the space and the floor joists to support one of these big guys, they are worth considering.

Wood Furnaces or Boilers

Wood furnaces and boilers come in two varieties: indoor and outdoor. Outdoor wood boilers are super large wood stoves, surrounded by a tank of water or antifreeze solution. Heat from the burning wood is transferred to the water, which is then pumped into the house for space and water heating. A control system located in the house regulates the temperature of the boiler, by dampening the fire. Most units also provide an alarm signal to indicate when it is time to re-stock the unit.

Figure 4-11. Outdoor wood fired boilers such as this model, keep the wood and chips outside.

Indoor wood boilers and furnaces are usually dual-fuel fired, such as the unit shown in Figure 4-12 from Benjamin Heating Products (www.benjaminheating.com). These units are very similar to outdoor models, except for their size and wood appetite. A major advantage of indoor, combination-fired boilers is the ability to supply heat, even after the fire has burned down. The Benjamin unit is both a high-efficiency wood-burning stove and oil-fired backup burner. The backup burner is only used when the wood fired section cannot supply sufficient heat to maintain the desired boiler temperature. Units are available with fan forced heating into a standard central air plenum or for use with hydronic heating, which is discussed later in this chapter.

Moving the Heat Around

The most popular style of heating system is the central warm air distribution design. A thermostat commands a centralized furnace to heat an air plenum or chamber. An electric fan circulates room air through the hot plenum and a series of ducts throughout the house. When the thermostat senses the room temperature is warm enough, the furnace turns off and the cycle repeats.

Figure 4-12. The combination boiler provides space and water heating in one compact unit. This model burns wood and oil, providing maximum flexibility in fueling options. (Benjamin Heating Products)

Hot water or steam boiler systems substitute water for air. These systems utilize either radiator units to transfer heat to the room or hydronic, in-floor hot water pipes as shown in Figure 4 - 17.

A wood stove or masonry unit uses direct radiation and convection to distribute its heat.

When determining your choice of heating system, consider the following:

• Will the renewable heating system be your primary heat source or will it be supplementary? If it is to be your primary heat source, you must consider backup-heating methods, particularly in colder areas, or if you will be away extensively.

• If you live in a more temperate area and travel little, and renewable heating is to be your primary heat source, consider pellet or masonry heaters. These models can provide sufficient heat for up to one-and-a-half days. Heating for longer periods of time will require a friendly neighbor to stock the stove, or an alternate fuel heat source with automatic controls that activate when the fire dies.

• If the system is being used off the electrical grid, energy consumption of the furnace, boiler, fans and control systems must be considered in relation to available electrical energy production.

The Specifics of Renewable Energy Heating

Anyone who has ever been to a cottage that is heated with an old Franklin or cook stove heater cannot imagine using wood as a primary source of heat. The old stove gobbles up armfuls of wood and never seems to get the place really warm. Other

Figure 4-13. A wood stove placed in a central location near open areas of the home is best. Remember to plan access for wood supplies coming in and ashes going out.

times, it's the opposite problem; there could be too much heat and belching smoke, making you wonder if your cottage has been magically transported straight into the depths of hell.

Many people use wood or wood pellet stoves as their primary heating source. If coupled with proper energy conserving techniques and passive or active solar heating, you will have the best heating synergies. The answer to avoiding the cottage nightmare is planning and using quality equipment.

You have several means of working with wood as a primary heating source:
- convection space heater
- wood fired furnace (single- or dual-fuel combinations)
- hydronic or hot water in-floor heating

Convection Space Heating

Convection space heating is the most common type of wood heating installation. No electricity is required to operate this type of unit, which is doubly good for off-grid applications. Locate the wood stove in a central part of the house, with open passages to the upstairs and other rooms. An example of a well-placed unit is shown in Figure 4-13, where heat from the stove can easily move around the home and up the open stairwell.

A common problem with space heating units is purchasing a model too large for your needs. Make sure you discuss sizes with your dealer, and explain your needs

Figure 4-14. A wood pellet stove offers the look of a traditional wood-burning stove with the convenience of only having to add a few scoops of pellets into a hopper to provide equivalent heating capabilities. (Harmen Ltd.)

and floor plan. Most first-time wood stove users cannot believe that such a small unit is capable of heating an entire house, but it will! On the other hand, if you are the sort who likes to wear shorts in January, maybe this isn't such a bad thing.

If your heating area is enclosed, or broken into many rooms, it is possible to use small "computer fan" units to blow air down hallways or between rooms. These units are available at wood stove dealers. Many building codes allow for grates to be cut between the ceiling of the heated room and the floor above, allowing natural air circulation.

Another option is to consider a pellet stove in tight locations. Because of the precise electronic control systems, it is possible to lower burning rates below a typical wood space heater and still provide a glowing hearth. An example of a Harmen Ltd. space-heating pellet stove is shown in Figure 4-14.

A wood pellet stove hopper holds up to one-and-a-half days worth of fuel. This offers the added advantage of not having to run back home to re-stock the stove or bugging your neighbor should you decide to stay out late.

Homes with an existing central furnace provide an automatic means of supplying backup heat. When the fire goes out, the thermostat turns on, kicking in the furnace. Homes thusly equipped also provide an alternate means of moving the

Figure 4-15. The air distribution provided by the central furnace fan provides the best method of moving heat away from the wood space heater. A ceiling fan is a good choice for cathedral ceilings or off-grid home.

heat from your wood stove around. Figure 4-15 shows such a home. Heat from the wood stove pools at the ceiling level. An existing or additional cold air return or suction duct is located near the ceiling. The furnace fan is left in "manual" or low-speed mode, drawing excess heat from the room and circulating it throughout the house.

An effective alternative for homes without centralized air distribution or having cathedral ceilings is to place ceiling fans in the heated area. Air circulation causes a mixing of the warm ceiling air with cooler room air by blowing in a downward direction.

An off-grid home has fewer choices for backup heat than those connected to the grid. What happens when you are away? Rather than missing the trip to the in-laws on New Year's day, install a propane fireplace or free-standing unit (high-efficiency of course) similar to the one shown in Figure 4-16. Not only do these modern units not require any electricity, they can be sized to heat any home. Besides, they are a lot nicer to look at than the old furnace in the basement. Propane is the secondary fuel of choice for off-grid homes. Although propane is not renewable, governments consider it one of the cleaner energy sources.

Wood Fired, Hot Air Furnaces

Wood fired furnaces operate in the same manner as fossil fueled units. The main advantage of wood furnaces is the fan forced circulation system that allows even heat distribution. All models contain automatic dampers, thermostatic fan control and safety shutdown dampers. Due to the air heating plenum design, central air conditioning may be added without difficulty.

Several models of wood-fired hot-air furnace, similar to the unit in Figure 4-12, are designed as dual-fuel

Figure 4-16. Propane fireplaces, radiant heaters and freestanding units such as this model require no electricity and can be sized to meet the heating requirements of any house. They also look a lot better than the $5,000 furnace in the basement.

units. These systems switch from wood to the secondary fuel source automatically as the fire dies down. This type of an installation is ideal where winter travel plans are expected.

Fan-forced hot-air furnaces are generally not recommended for off-grid applications. This recommendation stems from the energy requirements of the blower motor during the winter months when renewable energy electrical production is reduced. Possibly the most efficient fan-forced model is the Benjamin model similar to one shown in Figure 4-12. Its energy requirements are:

250 Watt motor x 7 hours per day = 1,750 Watt hours per day consumption

Depending on your location and heating requirements, you may need more or less fan running time. This simplified example shows you that even a unit with a very small, energy-efficient fan requires a large percentage of your electrical generation capacity off-grid. For you on-grid folks, this energy requirement would be considered quite low and is not a concern, except if power outages are a problem in your area.

Figure 4-17. Hydronic heating works by pumping hot water through plastic tubing located in or under your home flooring.

Hot Water or Hydronic Heating

Hydronic heating systems are becoming increasingly popular. Part of the popularity is continuous, even warmth, with all flooring materials including ceramic tile remaining warm and cozy. Many people who suffer from air borne allergens find that their symptoms are greatly reduced with hydronic heating, as there is no dusty air being blown about the house. Lacking a fan, hydronic systems are also very quiet and energy efficient.

The concept behind hydronic heating is quite simple. Heat water and pump it through a series of flexible plastic pipes located in the flooring of your house. A zone thermostat turns a very tiny, energy efficient pump on and off as heat is required in each zone. For areas where freezing is a problem, environmentally friendly anti-freeze is added to the water. This is also required for outdoor boiler heating systems described above.

Another advantage of hydronic heating is with the boiler unit itself. Most boiler units provide two "heating loops", allowing one loop for hydronic heating and the second loop for domestic hot water heating. Dual-fuel boilers can provide nearly 100% of your home space and water heating requirements using renewable resources. For those times when the fires burn out, the auxiliary fuel (preferably biodiesel) takes over the heating load.

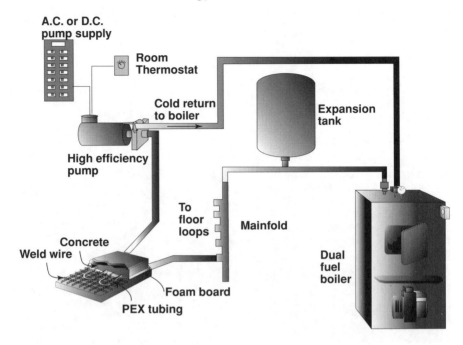

Figure 4-18. A hydronic heating system is shown.

Take a look a Figure 4-18 and follow through the operation of a typical hydronic space-heating system. A dual-fuel fired boiler provides the energy for both space heating and, optionally, a second plumbing circuit for domestic hot water. The dual-fuel fired boiler has a wood burning section on top and an oil-fired burner on the bottom. Provided the wood fire is maintained, the boiler is programmed not to start the oil burner unit. This allows you to supply nearly 100% of your heating needs from a renewable source.

Hot water or water/antifreeze mix leave the boiler and enter a manifold line. The manifold line contains an expansion tank that is placed in the system to allow the water to expand or contract in volume depending on temperature. The manifold provides a number of outlets to feed the various zones or area heating loops.

The heating loops comprise a flexible plastic pipe known as PEX, manufactured by such companies as Rehau Inc. (www.rehau-na.com). PEX pipe may be encased in cement flooring that is insulated over grade, or installed directly onto the bottom of interior sub floors. Heating an insulated concrete slap provides a high level of thermal mass and helps to regulate the temperature of the room. This provides much higher levels of comfort than central hot air furnace designs, where air temperature fluctuates over time. PEX lines should be no longer than about 300 feet (90 meters), as the heat dissipates with distance. Cold water exiting from the heating loop is drawn by a high-efficiency circulation pump and returned to the boiler water inlet. Hydronic heating contractors utilize Grundfos (www.grundfos.com) circulation pumps, which are rated for approximately 30 Watts at 120 Volts. For on-grid systems, these are one of the finest pumps and have been proven to operate reliably over many years. For off-grid systems, even a rating as low as 30 Watts is still too high. A better selection is the El-Sid (Electronic Static Impeller Drive) pump from Ivan Labs (www.ivanlabs.com). This pump does not have a conventional motor and spinning shaft, but uses an electronic drive circuit to cause a magnet embedded in an impeller to spin. Being completely sealed, it will last "forever". The El-Sid pump is also very efficient. A 10 Watt, 12 Volt DC pump can circulate water through two 300 foot (90 meter) heating loops. To put this into perspective, most people use 15 Watt CF lamps for each table lamp. The fact that it is a DC driven pump means that it can be directly connected to the battery bank or through a converter module, which will be discussed later in Chapter 5. Direct connection to the battery bank, saves electrical energy through inverter losses.

The controls for a hydronic heating system couldn't be easier. The pump is connected directly to the thermostat, which in turn powers the pump on and off based on heat demand.

Heat loss calculations, PEX tubing layout and other plumbing factors necessitate a discussion with your heating contractor to determine design and installation details.

Although I said earlier that active solar heating was beyond the scope of this book, a quick word right now would be prudent. A hydronic heating system coupled with solar heating panels, such as the Thermomax system

Figure 4-19. Active solar heating panels such as this model from Thermomax supplement your renewable energy hydronic heating system.

(www.thermomax.com) shown in Figure 4-19, provides excellent synergy. Using advanced differential thermometer controls, heat energy can be picked up by the solar collector and transmitted directly to the hydronic heating fluid, supplementing or replacing the boiler's energy. When the sun says "good night", heating fluid is stopped from flowing to the now cold solar collector. Companies such as Thermomax are available to provide needed engineering and component selection.

Wood Fuel – The Renewable Choice

Nature provides us with an endless supply of wood. Dead trees rotting on the forest floor produce the same amount of greenhouse gases as if you had burned the wood in your stove. Sustainable woodlot management requires the thinning and cutting of damaged and dying trees. Even in a small acreage, this "waste wood" will provide sufficient fuel for the largest house.

The species and quality of the wood you burn will have a major impact on the ease of use of your heating system. Freshly cut, wet wood can contain up to 50% moisture by weight. Attempting to burn this wet wood will result in difficult ignition, with reduced heat output and greatly increased pollution. On the other hand, properly dried and seasoned wood will ignite rapidly and provide nearly twice the warmth with less work.

Firewood should be cut and split (at least in half) early in the spring and properly stacked. Piles of wood should be neatly arranged in covered rows, allowing space between each successive row for air to blow through. The summer warmth and breezes will quickly reduce the weight of this wood by half, saving your back when you bring it inside next winter. During the seasoning process, wood will start to crack and split on the ends and turn a grayish color. Picking up two pieces of well-seasoned wood and banging them together will create a clear ringing tone. Doing this to freshly cut wood will result in a "thud".

The type of wood you purchase or cut will also make a difference. Softwoods such as White Birch and Poplar are fine for fall and spring when you are just a bit chilled. However, winter heating requires serious heating wood such as maple, oak, elm and ironwood. These woods have a higher density than softwoods resulting in higher heat output. Pick up an armful of White Birch and an armful of Maple. The Maple feels twice as heavy as the Birch and, guess what, it puts out about twice the energy for the same volume. As wood is sold and trucked by volume, purchasing the hardest woods will save you money and reduce the number of trips to the wood shed. Just remember to get at least a little mix of the softer woods for those winter days when the sun is providing some energy to the heating mix.

Purchasing Firewood

Your local dealer will offer you a confusing blend of softwood, mixed wood, mixed hardwood, fireplace cords, face cords and full cords, green or seasoned. It is important to understand what they are offering, otherwise your experience with heating in winter may make you think you are back in colonial Pennsylvania.

Figure 4-20. Firewood should be properly stacked and covered and be allowed to season for at least six months to drive off moisture. Dry firewood should be a grayish color and have cracks on the ends.

Firewood is sold in "face cords" or "full cords". The wood is typically delivered pre-cut in 16 inches (40 cm) long logs. When stacked in three rows, a full cord of wood should measure 4 feet deep x 4 feet high x 8 feet long. (1.2 x 1.2 x 2.4 m). A face cord is, as the name implies, one row or "face" of a full cord, which is equal to one-third of a full cord (16 inches deep x 4 feet high x 8 feet long). But be careful, if the logs of the face cord are cut into 12 inch (30 cm) lengths, it will yield only a quarter of a full cord.

Firewood sold "green" means that the wood is not seasoned. If you can get a discount for buying green firewood and have the time to season it, go right ahead. Wood can also be purchased in 4 or 8-foot lengths. Provided you have the tools, time and stamina to cut and split your own wood, the savings may be well worth the effort. To quote Thoreau, "Wood heats you twice, once when you cut it and again when you burn it".

How Much Wood Do I Need

The best way to figure this out is to try. A well-insulated house, such as our 3,300 square foot, ultra-insulated house uses 2 full cords (4 x 4 x 16 feet/ 1.2 x 1.2 x 4.8m) of mainly hardwood (and a bit of softwood) per year. We also travel during the winter and consume additional propane backup fuel when we are gone. A much smaller, turn of the century stone home just down the street, which has no insulation to speak of uses 4 full cords. It just depends. The best thing to do is buy more. Covered wood will not rot and should last up to 3 years, after which it will start to decay. So purchase a bit more and let it sit like money in the bank. You can always use it next year.

How To Cool A House

Cooling is another story. Central air conditioning is by far the largest and least efficient load in the home. It does not belong in an off-grid home. The best way to keep your home cool is to stop the heat from getting inside in the first place. This might sound simplistic, however, a well-insulated home, with shading that blocks the summer sun from entering the house will go a long way to prevent overheating. Open windows at night to create a cross-breeze and cool the house. In short, follow all of the guidelines discussed in *Chapter 3 – Energy Conservation*.

For those of you lucky enough to have relative humidity levels below 30% in the summer, an evaporative cooling unit may be the ticket. These devices are very common in the southwestern region of the U.S. and use very little electrical energy to boot.

When the mercury and humidity rise, even the best built house needs a little help. A one-ton (12,000 BTU) per 1,500 square foot, high-efficiency window-style air conditioner will help. In our house, one unit is installed on the main floor with a second installed in the bedroom area upstairs. Both units are permanently mounted into the walls, with the condenser (the part outside) facing as far north as possible, out of direct sunlight. Although these units are very large loads on the off-grid system, they are just the ticket to bring the humidity and temperature down to a comfortable level. Every A/C installer will swear up and

Figure 4-21A/B. Now you see it, now you don't. This one-ton (12,000 BTU) air conditioner is only required occasionally. Installing it permanently in a wall-mount design as shown here reduces the amount of encroachment on your living space and allows for easy hiding.

down that this size is too small. If you are the sort who needs to wear a nice wool cardigan in your house in July, then maybe they are right. For the energy conserver, this size works quite well.

Technically speaking, most A/C systems are sized far larger than needed. This is usually an effort to make sure that "you are getting what you paid for": a very fast, obvious cooling of the indoor air. It also ensures you don't call the installer back because the A/C isn't working well. On the other side of the coin, a large unit cools the air, but does not have sufficient time to reduce indoor humidity levels. A smaller unit running for a longer period will ensure lower indoor humidity and temperature. One big plus for these smaller A/C units is that they tend to be used only when there is a surplus of energy. This normally occurs on those long, hot summer days that make the PV panels so happy.

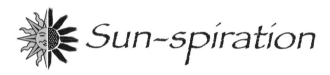

Sun-spiration

Prior to building their off grid home, Karen and Jamie Wilson placed their home in the suburbs for sale......

"It was supposed to be nothing more than a dry run. Start reducing our electrical loads in our grid-connected house to prevent going cold turkey in the new off grid one.

Compact florescent lamps, eliminate phantom loads, TV and stereo on wall switches, a new fridge. Bit by bit, the electrical bills started to drop until we reached the lowest consumption charge and minimum payment. Boy, what an easy way to save money.

Our minimalist energy bill came with a surprise end as well. A number of homes were for sale in the neighborhood and we were faced with competition. One couple looked at ours and just couldn't believe what an energy efficient house it was. They were so skeptical; they even checked the hydro bills to make sure everything was OK! We closed the deal shortly after."

Jamie Wilson

5.
A Renewable Energy Primer for On- and Off-Grid Living

I remember the day I popped the "big question" on Lorraine. The anticipation and excitement were now quickly concluding. She was clearly caught off guard. After a moment's pause came the reply, "But can I still use my hairdryer?"

That was her only concern. The rest I was sure, would be easy…

Nine years have passed and yes, Lorraine can still use her hairdryer. Living off-grid is both comfortable and satisfying. Our motivation for leaving "life on the grid" was simple. Lorraine wanted to move closer to her family and still have the room and privacy to support her "addiction" to animals. The lot at the back of the family farm fit the bill (and the wallet). There was only one downside: it was about $13,000 from the nearest hydro lines.

My work as an electrical/electronics hydropower engineer made me think: "Why not try to make our own juice?" Surely, I could whip up something for around $13,000. In hindsight, that was a bit naïve, but in the end, our system has grown to the point where it supports our lifestyle and is far more reliable than the power utility. Since that time I have shared my experience and knowledge with dozens of homeowners who have since taken the renewable energy off-grid or grid-interconnected plunge.

What is the trick? A willingness to give it a try and a touch of adventurousness sure, but most of all it's about energy conservation. The previous chapters have been ingraining this into your mind. Just in case you skipped all of that, let's take a moment to reiterate its importance.

Energy conserving on- or off-grid does not mean you have to live a Spartan lifestyle. A big screen TV, computers, and a cappuccino maker are examples of appliances and devices that are in our off-grid home. Add in the lights and stereo in the horse stable, a hot tub on the deck, garage with electric door openers and you might think that this house is a large electrical consumer. In fact, the opposite is true. We operate our house on between 3 to 6 kWh per day, depending on the season. Contrast this with other homes that use 40 to 75 kWh per day.

An energy efficient house has much of the same "stuff" as a regular house, but it uses 10 times *less* electricity than the average home. Before you run off saying, "Yes, but all the expensive appliances to operate use propane", remember that most people use natural gas for the majority of these loads, and dollar-for-dollar, electricity is the most expensive way to make heat. Ask someone who just paid his or her electric heating bill for this past winter.

The Fun Stuff – Making Electricity

We now know how to stretch our electrical energy dollar further and live within the means of an off-grid system. Let's see how to make the energy that we need to run a household. It should be obvious if we consume 3 to 6 kWh per day, we need to produce about that much too.

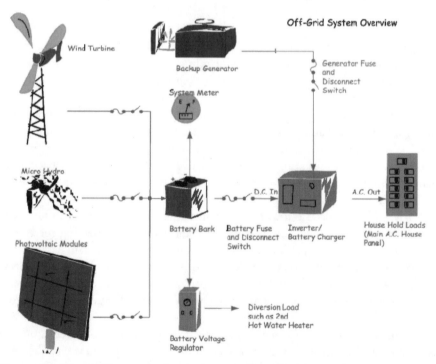

Figure 5-1. Renewable energy electrical systems for off-grid use rely on one or more renewable energy sources charging a storage battery bank. An inverter converts the energy stored in the battery to regular household power to operate your appliances. On-grid systems are similar, except that battery storage is optional, and the house remains connected to the grid.

Figure 5-2. Unlike fossil fuels or nuclear energy, renewable energy sources are just that, renewable. No matter how much sunlight we capture with our photovoltaic panels, there will always be more for everyone else; moreover, we're not dumping today's pollution on tomorrow's children. The photo above shows a 1200-Watt peak-output PV array on a sun tracker mount.

All off-grid systems work much in the same fashion. Collect energy from a renewable source (wind, sun, and/or stream), convert it to electricity and store the energy, usually in a battery bank. When we need this energy to power our appliances, we take some of the energy from the battery, convert it to alternating current (just like the power utility) and feed it to our unsuspecting appliances. Most off-grid systems also contain a backup power source, usually in the form of a propane, gasoline or diesel generator or *genset* for short. These units supply power to charge the battery bank in periods when the renewable system is not able to support the necessary electrical loads.

Figure 5-3. With advances in materials technology, you will find wind power used in every location on earth. The Bergey 1500 small wind turbine is specifically designed for home sized off-grid and grid-interconnected applications.

On-grid or grid-interconnected systems are very similar to off-grid systems. The renewable energy is fed directly to the inverter and converted into alternating current, which in turn supplies the grid. If you produce more energy than you require, the surplus is sold to the grid, so that your neighbour down the street can also use some. Selling energy causes your electrical meter to run backwards, providing you with a credit in your electrical energy "bank account". When you require more energy than you are producing, the electrical grid supplies you with "make up" energy causing your meter to debit your account in the normal manner. Every once in a while, your utility will send you a statement indicating whether you owe money or the present balance in your energy account is positive. The selling of electrical energy back and forth is called net metering. It is the law in many North American jurisdictions, requiring the utility to purchase your excess energy at the same retail price you pay for theirs. This really is a great deal.

Figure 5-4. An off-grid system requires a means of storing the electrical energy generated by the renewable sources for later consumption. A sizeable battery bank provides the needed storage.

Grid-interconnected systems do not require a battery bank and voltage regulation equipment, or backup generation. This lowers the installed cost of the equipment as compared to off-grid systems. On the other hand, if you have no batteries or backup generation source, your home will be just as dark as your neighbour's during the next electrical blackout.

In theory, it seems simple, but just like everything else in life, the devil is in the details. Look at the overview shown in Figure 5-1 and lets follow the system through its operation.

All of the earth's energy comes from the sun. In the case of renewable energy sources and how we harness that solar energy, the link is often very clear: sunlight shining through a window or on a solar heating panel creates warmth, and when it strikes a photovoltaic (or PV) panel the sunlight is converted directly into electricity; the sun's energy causes the winds to blow, which moves the blades of a wind turbine, causing a generator shaft to spin and produce electricity; the sun evaporates water and forms the clouds in the sky from which the water, in the form of raindrops, falls back to earth. The rain falling in the mountains becomes a stream that runs down hill into a micro hydroelectric generator.

While these energy sources are renewable, they are also variable and intermittent. In order to

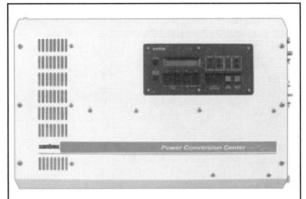

Figure 5-5. The modern off-grid world runs its electrical appliances on alternating current supplied by an inverter such as the one shown here. (Xantrex Technology Inc.)

ensure that electricity is available when we need it, a series of wire cables, fuses and disconnect switches delivers the energy to a battery storage bank.

Although there are many different types of storage batteries in use, the most common and reliable by far is the deep-cycle, lead-acid battery. You may be familiar with smaller ones used in golf carts or warehouse forklift trucks. Batteries allow you to store energy when there is a surplus and hand it out when you are a bit short. So why are we using a battery bank? What other means do we have to store electricity?

Great questions, simple answer: There is not any other feasible method of storing electrical energy. Maybe down the road, but if you want an off-grid system now, batteries are the only way to go. Today's industrial deep cycle batteries are a solid investment that should last 20 years with a minimum amount of care. At the end of their life, old batteries are recycled (giving you back a portion of their value) and new ones are installed.

Storing electrical energy is simple: just connect the renewable energy source to the battery and away it goes. Getting it out is a bit more complex. First, electricity is stored in a battery at a low voltage or *pressure*. You probably know that most of your household appliances use 120 Volts, whereas off-grid batteries commonly store electricity at 12, 24, or 48 Volts. The electricity stored in a battery is in a *direct current* (DC) form. This means that electricity flows *directly* from one terminal to the other terminal of the battery. Direct current loads and batteries are easily identified by a red "+" and black "-" symbol marked near the electrical terminals.

The electricity supplied by the utility to your home is *alternating current* (AC). This means the direction of flow on the supply wires changes direction at a rate of 60 cycles per second or 60 Hertz. (Many of the terms used in electricity are named after the early inventors who discovered the physics surrounding a term. James

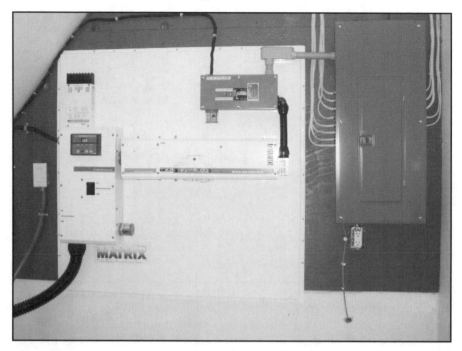

Figure 5-6. A renewable energy system should be neat, simple and well laid out, ensuring smooth sailing with electrical inspectors and your insurance salesman.

Watt, Count Volta, Heinrich Hertz, are a few of these experimenters).

In order to increase the voltage (pressure) of the electricity stored in the batteries and convert it from DC to AC, a device known as an *inverter* is required. Without an inverter, your choices in electrical appliances and lighting would be reduced to whatever 12 Volt appliances you could find at the local RV store. Early off-gridders did in fact choose this path, but do not consider it for anything but the smallest of summer cabins or camps. A house full of middle class dreams means a house full of 120 Volt, AC appliances. Standard electrical power also means standard wiring, standard electricians and happy electrical inspectors who enforce safety standards.

A supply of electrical energy from wind, water, or sunlight feeds low voltage electricity into a battery bank. The batteries store the electrical energy within the chemistry of the battery "cells". When an electrical load requires energy to operate, current flows from the battery and/or the renewable energy source at low voltage to the inverter. The inverter transforms the DC low voltage to AC higher voltage to feed the house electrical panel and the waiting appliances.

Solar Insolation Map
Average Hours of Sun for the Worst Month Yearly

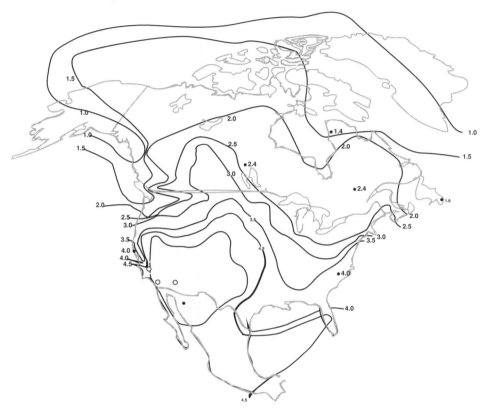

Figure 5-7. Sunlight hours per day are subject to extreme seasonal variability in most of North America.

The Changing Seasons

As a bad July sunburn will remind you, the amount of sunlight in summer is much greater than in winter. Simply put, the longer the sun's rays hit a PV panel, the more electricity the panel will push into the battery. The months of November and December tend to be dark and dreary by contrast. How does this affect the system and will there be enough energy in the winter?

Seasonal variability is extreme in the northeastern section of North America. The two maps in appendix 5 and 6 show the average amount of sun hours in the worst month and as a yearly average, respectively across North America. The amount of sunlight shining in December is approximately half as much as is shining in September, and even less than in June. Obviously, the PV panels' output reduces accordingly, and the amount of stored electricity varies with it. This creates an odd paradox. There is too much electricity in summer and not enough in November and December. How do we design around this problem?

Hybrids (Winter Season)

Hybrid design simply means adding more than one source into our energy mix. In our overview example, we have PV, wind and micro hydro, plus a backup generator. This design is not typical, as most off-grid systems typically start with PV as the main renewable source, a backup generator second and possibly a wind turbine third. For those of you lucky enough to have a year-round stream sufficient to operate a micro hydro system, that may be the only energy source you will require.

Grid-interconnected systems are typically PV based. Wind and water based sources are not commonly grid-connected, owing to the rural nature of these energy sources.

Back to watts and nuts and bolts for a second. Remember that we talked about consuming 3 to 6 kWh of electricity per day. Our PV panels' rating is 1200 Watts peak-power output (28 Volts x 43 Amps DC). In reality, they tend to output approximately 950 Watts under ideal conditions, less if it is hazy and nearly zero if the day is cloudy. The entire assembly is mounted on a sun tracker unit, which allows the panels to face the sun as it moves from early morning through late afternoon, winter and summer.

Referring to the worst month sun hours map in appendix 5 for our location, we find our average to be 2.2 sun hours per day:

2.2 sun hours per day x 950 Watts output = 2,090 Watt-hours per day or approximately 2 kWh per day

With 2 kWh of production and an average consumption of 4 kWh per day, the system will loose 2,000 Watt-hours per day. If this was your bank account and you kept taking out more money than you put in, guess what happens? Depending on how deep your pockets are, you run out of cash. The off-grid system batteries are no different. In fact, normal battery sizing assumes that you should be able to run your house "normally" for 3 to 5 days without having *any* input from your renewable energy sources. For our household, running average loads means the batteries need to supply:

5 days supply x 4 kilowatt-hours per day = 20 kilowatt-hours usable capacity

So, what happens at the end of 4 days? This is where the hybrid design comes in, you either have another renewable source pickup some of the load or rely on your backup genset. Depending on the degree of automation in your system, you either manually start the backup generator (gas, diesel or propane) or a generator control device starts the generator for you. In either case, the inverter now switches to battery charging mode and fills the batteries back up. The house electrical loads automatically receive power from the generator during this charging time. Once the batteries have reached full charge, the generator turns off automatically or you run out in your housecoat and slippers to shut it down. (I think the automatic feature is definitely worth the few extra bucks!)

If your system contains more than one renewable source, you will find that they tend to be complementary. A dull day in November often has brisk winds and, conversely, the air on a sunny summer day is hot, still and stifling. However, do not believe that having PV and wind will eliminate the need for a backup generator; it will not. The combination will reduce the running hours of the generator considerably, but it will not eliminate its necessity. Our house still requires over 100 hours of generator time per year.

Figure 5-8. This large 7.5 kW diesel backup generator set (or genset) will last several lifetimes and be economical to operate as well.

Hybrids (Summer Season)

During the summer months, the increase in sunlight hours, coupled with a lower need for lighting and less time spent indoors creates a surplus of energy based on consumption levels at 3.5 kWh per day.

Production:

6.0 sun hours per day x 950 Watts output = 5,700 Watt-hours per day or approximately 6 kWh per day

Surplus:

6 kWh/day produced – 3.5 kWh/day used = 2.5 kWh/day surplus

We may or may not need this surplus, depending on whether or not we require any air conditioning that day. As mentioned earlier, an air conditioning unit uses an enormous amount of energy, approximately 1,100 Watts per hour operated. Based on a surplus of 2.5kWhr/day, we should be able to operate the air conditioner for approximately 2.5 hours per day, without dipping into the energy bank.

On days that we do not need air conditioning, the surplus energy produced must go somewhere. You must consume this energy or the batteries would reach a fully charged state and would eventually "overcharge". To prevent this from happening, a battery voltage regulator connects to a diversion load. A typical diversion load consists of an electric water heater, plumbed "before" the regular gas water heater (see Figure 10-8). Converting the electric water heater to the same voltage rating as your battery bank makes the diversion load (24 Volts in our home).

While operating, the battery voltage regulator monitors the battery voltage or state of charge. When the batteries become fully charged, the regulator starts to divert surplus electricity to the electric water heater. The water starts to heat as it absorbs the extra energy produced by the renewable energy sources. Over the course of a day or two, the water can easily reach 140 degrees F. (60 C.), which in turn flows into the regular gas water heater. As the incoming water is already hot, the gas heater remains on standby, thus conserving propane gas, energy dollars and the environment; energy is never wasted in this system!

Phantom Loads

We discussed phantom loads briefly in Chapter 3 – Energy Conservation. As the name implies, phantom loads are any electrical loads that are not doing immediate work for you. This includes items such as doorbells (did you know that your door bell is always turned on, waiting for someone to push the button?), "instant on" televisions with remote controls, clock radios, and power adapters.

So, what is the big deal? First, these devices are consuming energy without doing anything for you. That electric toothbrush was probably charged about 15 minutes after you put it back in the holder. The remaining 23 hours and 45 minutes until the next time you use it represents wasted energy. A television set that uses a remote control is actually "mostly on" all the time, just waiting to receive the "on" command from your remote.

While the total dollar cost for these luxuries is small on-grid (around $5.00 per month for an average house), this consumption off-grid is unacceptable. By the way, this "small" bit of waste equals quite a few million dollars a month in North America alone.

Another reason for the concern about phantom loads in off-grid installations is the inverter unit waste. This device goes into a sleep mode when the last light is turned off at night, conserving a fair amount more energy. Any phantom load keeps the inverter "awake" consuming more energy than it otherwise would.

Figure 5-9A/B. Phantom loads are unnoticed and often useless electrical loads that burn holes in your energy conservation plan. Examples are doorbells, "instant on" electronics and microwave ovens or anything that uses a "power adapter", like the one shown above.

Phantom Load Management

Some phantom loads can simply be eliminated. Try a doorknocker instead of a doorbell and a battery-powered digital clock instead of the plug-in model. I have even heard that some people use manual toothbrushes.

Television sets and CD, DVD, and VHS players with instant on and remote control functions should be wired to outlets that can be switched off. This could mean having the electrician wire in the switch for you or using a power bar with an integral switch.

Many people cannot live without some phantom loads such as a fax machine, cordless phone, cell phone, PDA and laptop chargers. For these items, use a separate wiring circuit run through the house, which connects to specially marked outlets. These outlets are reserved for ESSENTIAL "always on" loads like the ones described above.

The power for these outlets comes from an inexpensive, 100-Watt inverter that is wired directly to the batteries. Such an

Figure 5-10. Wire a small inverter such as this one to special outlets strategically placed throughout the house for those very small, "always on" loads.

arrangement ensures the main inverter can go to sleep at night and still provide power for the small devices you desire. If your home is already built, it may be possible to group all of these special loads at one central location and run the inverter to a power bar at that point.

Take note about this "small" inverter powering phantom loads. If you load it up with all of your toys, it is possible to burn up nearly 100 Watts of power. Over the course of 24 hours, this is a lot of juice:

100 Watts x 24 hours per day = 2,400 Watt-hours or 2.4kWh

A load of 2.4 kWh is almost half of most renewable-energy systems' daily total energy production; tread lightly!

Metering and Such

At this point you are probably wondering if I run around the house with a note pad and calculator, chastising Lorraine for using her hair dryer too long or making the toast too dark, while recording every volt and who knows watt. Actually, we hardly look at the system at all. Once you install your system and load the house with all the electrical goodies imaginable, within your average production limits, the system will almost take care of itself. If we use more energy than we produce, the generator may run for a while. If we make more energy than we use, the next shower is free due to the savings in hot water. The system is almost invisible.

For grid-interconnected designs, the system is invisible. The only notification you get that all is well is a statement from the utility advising you that you have a $500.00 credit.

As with any piece of complex machinery, a bit of care and management is required. I would include a multi-function meter in the mix. This unit monitors the energy produced and consumed, and calculates all the nasty mathematics necessary to tell you how much juice is really in the battery. By monitoring the meter and other data, you will have a comprehensive snapshot of the health of your own power station.

Figure 5-11. Every off-grid power system should have metering capabilities. A proper meter leads to increased load control knowledge and system reliability (Bogart Engineering).

Is it Economical?

This is a tough question, because the answer is not straightforward. Let's start by looking at an off-grid system. The cost of a turnkey (i.e. you do none of the installation work), off-grid PV-based system running house loads similar to ours is about

$12,000 to $18,000 for all of the materials and installation labour. Do some of the work yourself and it can be lower. Add a wind turbine and, of course, the cost increases. Can you work with used equipment and do you like to tinker around? Can you live with smaller or fewer electrical loads? Swap a 15 inch TV for the big screen, use regular telephones, charge the cell phone in the car, there are a million ways to lower your loads and system costs. If you do not do any energy conservation, expect to pay between five and ten times the amounts discussed. (Once again, energy conservation is important!)

How far the installation is from the hydro grid determines the break-even point for most off-grid homeowners. If your hydro utility is more than a half mile (0.8 km) from your house, it may pay for itself from the second you turn on the first light. You can also add to that the benefits of no hydro bill, zero environmental pollution and the feeling of self-sufficiency you get the next time your neighbour's lights go out during that dark and stormy night.

Will your private power system remain connected to the electrical grid? If so, the installed rate will tend to be lower, because of fewer equipment requirements. Remember that you will not require a battery bank, voltage regulator, battery compartment and many other components. On the other hand, most grid-interconnected systems tend to be larger than off-grid units. What leads to this apparent contradiction?

Connecting a renewable energy system to the grid is not currently economical at all, environmental issues aside. Why bother hooking up expensive equipment to the grid, when the payback period is dozens of years, if ever. Most people are not so committed to conservation that they are willing to spend loads of cash just to help the environment. The answer is net metering and subsidized paybacks in the form of capital equipment rebates. We discussed net metering above. This program allows you to sell power back to the utility at the same retail price you pay. Net metering laws are springing up all over the planet and for good reason. Electrical power from centralized fossil fuel plants is becoming an albatross that needs removing from society's neck. Power brownouts in California, grid disruption along the north east coast and damaging ice storms are all reasons why governments wish to encourage distributed electrical generation. Never mind the environmental concerns which speak for themselves.

At the time of writing this Handbook, California has the best support program to encourage renewable energy, distributed generation of electricity. The program is the "Emerging Renewables Buy down Program", which provides a cash rebate for up to 50% of the cost of these systems. One typical system having 2,500 Watts of PV, grid-interconnected, after rebate costs $7,500.00. Payback periods with these rates are not much different from typical energy efficient appliances.

Use care when calculating system payback. Do not assume you can use your current rate of electrical consumption in the payback equation. Just because your home is grid-interconnected does not mean that you should forego energy conservation. As discussed in *Chapter 3 – Energy Conservation*, it is better to spend one dollar on high-efficiency equipment to reduce loads than it is to spend five dollars on renewable energy production. The payback on renewable generation equipment will look worse once you have reduced your electrical loads by 5 or 10 times.

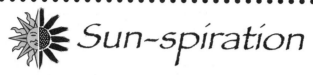

Sun-spiration

Lorraine's mother Mary Thompson had some recollections of life off the grid....

"When my daughter and son-in-law told me they were going to build a house and make their own power, it brought back some memories of growing up on a farm in rural Ontario.

We did not have electricity until approximately 1925. We were the only home in the area that had lights in our house for quite some time. We started with oil lamps that were set on shelves in case of fire and had poor lighting for doing homework. A hand washing machine, churning butter and separating cream by hand says it all! We then went to a Delco system that let us do some of the work with electrical power.

Our system wasn't hooked to hydro - that wasn't available yet. We ran on 24 batteries. There were two sets of batteries, with a ball on the side of the battery bank that dropped when it was time to switch to the other battery bank. When this happened, you had to start a generator to charge the first bank, while you used the second one. I was around 8 to 10 years old and was able to do this chore when needed.

We had a radio that was battery operated that you had to have a license to listen to. We were the envy of many and felt very lucky that my parents were able to provide us with such a luxury.

To get hydro into my home where I live now we had to purchase poles after waiting for several years for it to come close enough to us. It was set up that the hydro company had to have a number of properties per mile of country road to sign and pay for contracts. We had to sign three extra contracts just to get the line runs to us. Needless to say, six months later all who hadn't signed up to get hydro initially suddenly wanted it.

So when my daughter and son-in-law told me of their dream, I knew it wasn't crazy at all".

Mary Thompson

6.
Photovoltaic Electricity Generation

The process of capturing and using solar energy is as old as time. Like everything else in the modern world, the simplicity of capturing the sun's energy has been elevated to new technological heights. The photovoltaic cell used in renewable energy systems is definitely the product of rocket scientists. Unlike solar heating collectors that are used to warm fluids running within the collector, photovoltaic cells convert light from the sun directly into electricity.

The term photovoltaic is derived from the Greek language "photo", meaning light, and "voltaic", commonly referred to as voltage which assists the flow of electricity. Friends simply call them "PV" cells for short. Bell Laboratories discovered the PV cell effect in the 1950s. It didn't take the folks at NASA very long to figure out that PV cells would be

Figure 6-1. Photovoltaic cells are the backbone of renewable energy systems for homeowners. Since prices and supporting technologies have improved, hundreds of thousands of homeowners are now living lightly on the planet.

an ideal means of producing electricity in space. Many space missions later, PV cells have improved in performance and come down to earth in price. Nowadays, PV technology is used in watches, calculators, street signs and renewable energy systems for homeowners.

What is Watt?

If you look at the market for PV products, you will find a bewildering array of product types. For the home renewable energy system, PV products are relatively standardized allowing even a novice to make accurate comparisons between product lines. There are currently only three product technologies that should be seriously considered for home use: single crystalline, polycrystalline and string ribbon cell. Other cell technologies such as thin film or "PV roof shingles" may be considered, provided product warranty and manufacturer financial strength to honour the warranty period are acceptable.

Figure 6-2. PV cells are manufactured using a process similar to transistors, except they're larger. Individual cells are polished, interconnected and mounted in a frame creating a PV module.

When comparing module power ratings and performing home energy calculations, use caution. Manufacturer power ratings are based on ideal sunshine conditions, which rarely occur in the real world. It is wise to derate module power ratings by 40%.

Manufacturer module rating x 0.6 = "Real Power Rating"

...OR...

Real Power Rating ÷ 0.6 = Manufacturer Module Rating.

Discussing the cell technology is also a good starting point to understand how PV cells work.

PV Cell Construction

PV cells are transistors or integrated circuits on steroids. Most people have seen the latest microcomputer chip used in PCs. It's a silicon wafer about the size of your thumbnail that holds several million transistors and other electronic parts. PV cells start out the same way as chip circuits, but are kept in the oven and grown much larger, to approximately 4 inches (10 cm) in diameter. The baked silicon rods are then sliced into thin wafers. The wafers are then polished and assembled with inter-connecting electrical wires.

If we were to take one wafer, expose it to bright sunlight and connect it to an electrical meter, we would measure 0.6 Volts DC (Vdc) of electrical pressure. A voltage higher than the battery rating is required to charge the battery. For example, to charge a 12 Vdc battery, at least 15 Vdc is required, plus some additional voltage for electrical losses in the system. For this reason, PV cell manufacturers typically connect 36 cells in series to create an additive voltage. (Maybe you should consider reading Chapter 2 now if you haven't already.) A grouping of PV cells thus arranged and mounted in a frame is known as a module.

36 cells in series x 0.6 Vdc per cell = 21.6 Volts Open Circuit

This voltage appears to be a bit higher than our target voltage of just over 15 V. An interesting phenomenon of PV cells (or any producer of voltage) is a reduction of voltage when the cell is under load, for example when charging a battery. In addition to this, heat also causes PV cell voltage to drop. When the voltage of a PV cell or series of cells is measured without a connected load, we call this the "open circuit" voltage. Manufacturers often refer to this as "Voc".

When the module is at its maximum rated power output, the voltage is re-duced from the open circuit rating. This is known as the "voltage at maximum load", and is typically 17 Vdc for a 12 V unit.

In order to complete an electric circuit, we must have a source of voltage and current flow. Accordingly, a PV module will cause current to flow in the cells and out of the supply wires to the connected load. As an example, a Siemens (www.siemenssolar.com) model SP75 has a rated current of 4.4 Amps (A) at a rated voltage of 17.0 Vdc under ideal conditions.

17.0 Vdc x 4.4 A current flow = 74.8 Watts (W) of output power

This module is therefore rated at 75 W. There is one small exception: if it's dark or very cloudy outside, the power output is zero. Obviously light intensity has to play into our equation. PV manufacturers have taken the liberty of helping us out. The standard light intensity should naturally be that of the sun as this is the source of energy for our panels. But where should we measure our sunlight. In Alaska in winter, or perhaps the Bahamas in July? Lighting energy has been standardized by world agreement so that all manufacturers are using the same lighting intensity to prepare their data tables. Without this standardization, it would be impossible to determine which PV panel would be the best for your application.

Our sun has very generously decided to output a nice round 1,000 W of en-ergy per square meter at noon on a clear day at sea level. Light sources have been developed that create this same level of light intensity or *flux*. The light is beamed onto the subject PV panel, and its electrical ratings are recorded in the product data sheets. As a consumer, this greatly helps determine the relative value of one PV panel to another.

Figure 6-3. PV systems are upgradeable. This homeowner installed new single crystalline modules to increase the capacity of the original polycrystalline units located in the center of this array.

Cells can be manufactured using single, polycrystalline or string ribbon techniques. Single crystalline technology is more expensive and yields slightly higher light-to-electricity conversion than polycrystalline or string ribbon. Does it make any difference in the end product? Not really; all three technologies will provide a long and trouble-free life. Amorphous or thin-film technology should not be considered as long-term aging of this technology has not been proven in home applications. Due to the lower cost of amorphous technology, smaller units are being sold successfully into the RV market. Time, along with product reliability data, will determine if they are right for the home market in the future.

What is the Life Span of PV Modules?

Almost all PV module manufacturers provide a written guarantee for 20 to 25 years. The manufacturers are obviously quite sure that their products will stand the test of time. The reason for this certainty is the same reason old transistor radios last so long. The semiconductor technology of the cell wafer results in very little wear and degradation.

The standard warranty term from Siemens states that any module that loses more than 10% power output within 10 years will be repaired or replaced. In addition, if the same module fails to produce at least 80% of its rating after 25 years, it will be repaired or replaced.

The cells themselves are quite fragile. To protect them from damage and weather, the cells are bonded to a special tempered glass surface and sealed using a strong plastic backing material. The entire module is inserted into an aluminum, non-corroding housing forming the finished assembly. Once a grouping of modules, called an array, are mounted to a roof, fixed or tracking rack, they should stay put as long as you want them to.

PV Module Maintenance

Let it rain. That pretty well summarizes what you need to do to maintain your PV array. In the wintertime, ice and snow may build up on the glass. Do not smash the ice to remove this covering, unless you want to smash the glass too. In our area, a quick brushing with a squeegee will take off the loose highly reflective white snow. Once this coating is removed, the sun will quickly warm the panels even at −14 °F (-26 °C).

PV Module Installation Checklist

Place your PV module in the sun and collect power…that's it, that's all? Well, not quite. Although PV modules are well designed and last a long time there are several issues that must be considered before choosing where and how to install them.

1. Calculate your electrical generation requirements.
2. Ensure that the site has clear access to the sun.
3. Decide whether to rack or track.
4. Tree shading must not be allowed.
5. Consider snow and ice buildup in winter.
6. Ensure modules do not overheat in summer.
7. Decide on the system voltage.
8. Locate the array as close to the batteries or inverter as possible.

Step 1 - Calculate your Energy Requirements

This is square one on your quest for renewable energy. Let's start the ball rolling with a question: How much energy do you need? Not sure? Then go back to *Chapter 3 – Energy Conservation*. Energy conservation and load calculations are the most important part of the system sizing exercise. Remember that for every dollar of electricity you conserve, the cost of your system will go down by up to five dollars. You cannot afford to sidestep the conservation process, sorry!

As there is no standard or typical system sizing, for the purpose of the examples provided in this chapter, let's use an average electrical energy consumption of 4,000 Watt-hours (4 kWh) per day. To give you a bit of perspective on this value, Lorraine and I run our household on between 3 to 6 kWh per day. We use less if we are traveling, and more if we have friends over for a late night party. If you have teenagers that stay up all night, maybe your house should be grid-interconnected. To define "typical" is pretty hard, but here are some sample values that I can share from other installations:

- For a weekend cottage running a few lights, small TV, boom-box stereo and no refrigeration, a 12 Vdc system along with 2 to 4 PV modules and a small battery bank would suffice. 200 W peak power and 1,000 Watt-hours per day energy.
- For a seasonal cottage that is used full-time by four people, requiring refrigera-

tion and limited water pumping, a 12 Vdc system, with perhaps a mix of direct current and alternating current outlets for various loads, would suffice. If the cottage is to expand over time, wire for 120 Volts AC (Vac) now. Use a small inverter for 1,500 W peak power and up to 2,000 Watt-hours per day of energy. This will require a PV array of approximately 500-Watts peak output.

- For a full-time residence with four people who are energy wise, have no dishwasher or clothes dryer, a small television and stereo, high-efficiency clothes washer and 120 V refrigerator. Wire the system for 24 V batteries and provide an inverter with 2,500 W peak power and up to 3,000 Watt-hours per day energy. This requires a PV array of between 800 and 1,000 W peak output with a backup generator for "dark months". Depending on the location, more PV panels may offset the need for a generator, provided winter sun hours are able to provide 100% of the daily load.

- For a full-time residence with all the electrical (high-efficiency) goodies, may include a home theater system, dishwasher, gas clothes dryer, central vacuum, washing machine and electric refrigeration. The system may be wired for either 24 V or 48 V battery. Use 24 V for total daily energy consumption below 7,000 Watt-hours per day and 48 V for larger systems. Provide an inverter of between 2,500 W minimum and 4,000 W peak output or larger, depending on the final peak loading. This will require a PV array of 2,000 W peak or higher, depending on load. For off-grid systems, a backup generator will be required.

In Chapter 5, we discussed the variability and intermittent nature of sunlight. The amount of sunlight we receive varies from day to day, depending on clouds and rain, but also as a function of the seasons. In most locations, there will be a surplus of sunlight and resulting electrical energy production in the summer, with the opposite being true in the winter. It may be tempting to calculate the energy production of the PV panels using the worst sunlight hours of the year in an attempt to cover all of your energy requirements. Let's take a look at how this would work: assume you live in the rural Rochester, New York area. Turn to Appendix 5, *Solar Illumination Map for North America (worst months)*. Find the Rochester area and note that the amount of solar illumination in sunlight hours per day for the worst months of the year is 1.6 hours. We can now plug in some numbers to calculate the size of PV array we require:

4,000 Watt-hours/day ÷ 1.6 sun hours/day = 2,500 W PV Panel

Don't forget to derate the manufacturer ratings:

2,500 real power rating ÷ 0.6 = 4,160 W manufacturer rating

A quick search of the web on solar panels and guess what, you'll find nothing even closely approaching that figure. The largest 12 V module you can find from AstroPower produces around 120 W and has a list price of $685.00. You dig a bit further and find that Kyocera has an 80 W model on sale for $375.00. This is starting to get complicated and expensive.

4 PV modules connected in parallel

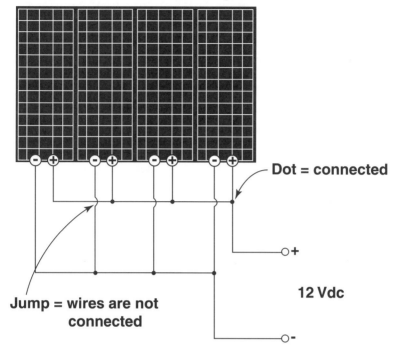

Figure 6-4. PV modules may be connected in parallel to increase wattage in the same manner as two smaller batteries connected in parallel increase their capacity.

PV modules are not any different than batteries in the way they are connected and used. Having read Chapter 2, you will recall that a grouping of batteries may be connected in *series* and *parallel* to increase the voltage or current, the same is true with our PV cells. If we desire a PV array of 12 V output, we can group our PV modules in parallel to increase the current and wattage rating.

4160 W Array ÷ 120 W PV module = 35 modules
Or
4160 W Array ÷ 80 W PV modules = 52 modules

With so little sunlight in the Rochester area in winter, this PV array will require a lot of modules to support 100% of the electrical house loads. Now let's take a look at the cost.

35 – AstroPower 120 Watt modules x $685.00 = $23,975.00
52 - Kyocera 80 Watt modules x $375.00 = $19,500.00

Although these figures are not exactly accurate due to changing prices in the market, it does show a surprising tendency. You would expect since each PV array makes the same amount of power that the cost would be the same. Welcome to the world of retail sales. This is no different than purchasing a car; it's best to shop around. I will digress a moment from our sizing example to explain how to comparison-shop for PV modules. If you are the sort who loathes people who drive a few miles to buy tomatoes at another store because they are 10 cents less, just wait a minute. PV module comparison-shopping is easy and will save you more than enough money to pay for this book and possibly put your kids through school.

The energy-to-cost ratio allows for a very quick comparison of all module types. The calculation is quite simple: divide the cost of the module by the rated wattage. We will use the example PV modules from the example above.

> *AstroPower 120 W modules @ $685.00 ÷ 120 W = $5.71 per watt*
> *Kyocera 75 W modules @ $375.00 ÷ 80 W = $4.69 per watt*

Assuming that the warranty and cost to install the additional modules required to make the same amount of power was taken into account, the Kyocera modules would be the least expensive. Now back to Rochester.

It takes quite a few modules to make 100% of the energy required to power a home when the winter sun hours are low. By contrast, the opposite will occur in the summer time. Referring to Appendix 6 *Solar Illumination Map for North America (yearly average)*. You will find that the number of sun hours per day has nearly doubled. In this example, the effect is obvious: for the time of year when the sun hours per day are peaking, the required number of PV modules in your array will be half the number required at the worst time of the year.

For people who are using their PV modules in a summer cabin or cottage, you are in luck. Use the calculations with the best sun hours per day. If you visit the cottage for brief visits in the fall and winter, possibly add another panel or two.

For year-round homeowners, the choices are a bit more complex. You can spend the extra dollars and purchase all of the panels you require, or consider a complementary renewable resource such as wind power generation. A third option is also available which tends to be the most common solution: add a backup propane, gasoline or diesel generator to help boost production at the worst of times. We will review all of these alternate options in greater detail in later chapters.

The correct number of panels to make up your array depends on more than just simply spending the money to purchase all you need. For starters, the weather seems to have a mind of its own. One year it's sunny all winter, the next you wonder if it will ever shine again. Charts like those shown in Appendix 5 and 6 are averages, which really means be prepared for variability. Even if you purchase all the PV modules you can possibly mount on your roof, it would still be a good idea to have a backup generator or alternate source just in case. A thousand PV panels will not light an energy efficient CF lamp if it's cloudy and dark!

Speaking of mounting on your roof, this is the next concern with the number of panels you select. You have several choices for where to mount your PV array. The roof is one of these places. The more panels you have in the array, the larger and heavier it becomes. This may limit some of the best mounting locations completely. If the panels are mounted on a sun-tracking unit, 16 to 20 modules is the

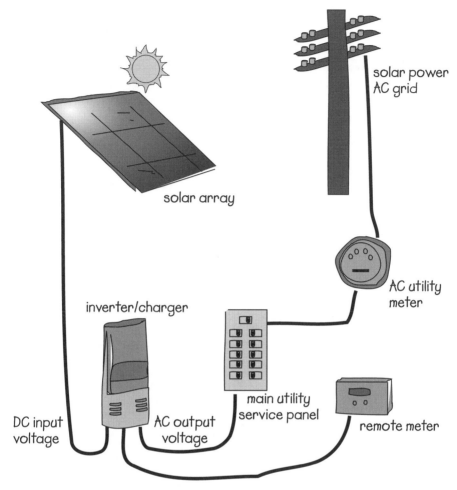

solar power
AC grid

solar array

AC utility
meter

inverter/charger

DC input
voltage

AC output
voltage

main utility
service panel

remote meter

Figure 6-5. Grid-interconnected renewable energy systems will make an impact on your electrical bill and the environment, regardless of its capacity. Most designs try to balance your daily consumption with generation. (Xantrex)

limit per tracker.

What is the correct number of panels to purchase? There is no "correct" answer. If you have deep pockets, by all means purchase all the modules you require for the darkest winter days. If you are on a tighter budget like most people, then start small and expand your system later as you can afford it. There is nothing wrong with installing 6 panels now and using a bit more generator power for the time being. When you win the lottery or your great aunt leaves you some Microsoft stock, splurge and purchase the additional 12 modules that you really need. The additional panels can always be added later, just like those shown in Figure 6-3. Just remember to plan for the expansion at the start of the installation. This will lower costs and future headaches.

Planning a grid-interconnected system? Even one panel will make a bit of difference. The government subsidizes most grid-interconnected systems, so it makes sense to install an array that will offset your home's electrical consumption. Do not get carried away with panels covering your property, the utility will not write a check for your contribution to supporting the grid. Most renewable energy grid-interconnected systems work on the net billing method discussed in Chapter 5. If you are planning the "go slow" approach by adding a few panels to the grid at a time, the only issue will be electrical safety inspection of your work. The utility charges a fee for each visit, so be sure to discuss this with your inspector before planning to add a 50-W panel to the grid.

Step 2 – Ensure the Site Has Clear Access to the Sun

PV modules must face the sun to make electricity. As every romantic knows, the sun sets in a brilliant display of color to the west. My mother-in-law tells me that it rises in the east, although I cannot personally attest to this statement. So where then do we face the modules? The quick answer is to point the panels directly at the sun. With the sun obviously moving from east to west during the day, and moving progressively higher in the sky as summer approaches, this might be rather difficult.

You have two options: mount the modules to a fixed location pointing solar south or install a tracking unit that automatically aims the panels directly at the sun.

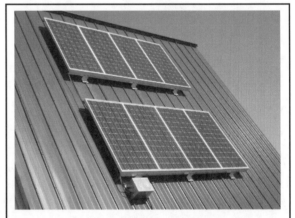

Step 3 – Decide Whether to Rack or Track

Roof Mounts

Perhaps the simplest mounting system is attaching the panels to a solar

Figure 6-6. Attaching your PV array to a roof facing solar south is one of the least expensive methods. The downside can be getting on the roof to brush off a foot of snow in the middle of winter.

south-facing roof as shown in Figure 6-6. Solar south differs from magnetic south due to a phenomenon known as magnetic declination. Appendix 4 shows the correction of compass readings based on your geographic location. If you live in Florida for example, there is no change between magnetic and solar south. If you live in Alberta, the error in compass to solar south reading is so large that it may seriously effect your system's energy production when using fixed mount PV arrays.

Another problem with fixed mount arrays is their inability to change their angle of view throughout the seasons. The winter sun barely scrapes across the horizon in most of Canada. During the summer, the change in the earth's angle places the sun almost directly overhead. If your roof or other fixed mount does not have the ability to be adjusted for summer to winter sun angles, a good rule of thumb is to set the array at an angle from the ground equal to your latitude. Mounting racks, which have a summer/ winter angle adjustment, should be lowered by 15 degrees for summer. Likewise, raise the angle by 15 degrees for winter.

Figure 6-7. The commercial ground-mounting unit is ideal for smaller PV arrays and where cost is a concern. (Zomeworks)

Ground Mounts

If the thought of climbing onto your roof to wipe snow off the PV array has you going woozy, then ground mounting is for you. A ground mounted PV array is the next simplest and lowest cost method of mounting the panels. They may be purchased, like the unit shown in Figure 6-7, or you can easily build a unit yourself. Select galvanized steel or aluminum to limit corrosion. When designing your mounting system make sure to include a hinge assembly to allow for seasonal adjustment.

A word of caution regarding ground mounting: PV modules are expensive and may have a tendency to walk away. Ensure that some form of security bolts or guard dog is used to prevent theft. Snow is another problem to deal with. If you live in an area where sweeping the array requires a snow blower, then ground mounting is not for you. Be careful to ensure that lawnmowers and playing children will not send rocks or other debris flying at the modules. Although it is very likely that no damage will result, there is no sense tempting fate.

Tracking Mounts

A tracking mount, as shown in Figure 6-8, is the most advanced means of tracking the sun and is almost hypnotic to watch it scan across the sky. PV arrays comprising up to 20 panels are mounted on the tracker, forming a billboard sized unit. (A 16-module array measures 16 x 8 feet/ 4.9 x 2.5 m.) The tracker is designed to move to an easterly location at dark and wait until sunrise, at which point it follows the sun

Figure 6-8. Tracking units such as these increase summer electrical production by up to 50%

on its westerly track across the sky. At day's end, the unit will return to the easterly heading to repeat the process. Trackers may be fitted with a manual seasonal adjustment device, or on the Wattsun unit (www.wattsun.com), an automatic control is available.

Passively controlled trackers are also available. The Track Rack manufactured by Zomeworks (www.zomeworks.com) uses heat from the sun to operate a tracking mechanism. While simpler in complexity than active tracking units, the Track Rack model will go to sleep facing west and may require an hour or two of sunlight before it will start correctly tracking on cold days.

Should you use a tracker? The debate rages on, but the following are considerations that may affect your decision to use one.

- Trackers increase summer PV production by up to 50%. Winter production is improved only 10 to 20% due to the lower, smaller arc of the sun.
- The further north you are the less sense it makes to track in the wintertime. This is especially true along the United States/Canada border area.
- If your summertime consumption increases due to additional loading such as swimming pools, hot tubs or air conditioning, increasing summer PV production with a tracker will help.
- If your site has a limited window of sunlight—less than 7 hours—then tracking will not greatly improve summer production.
- Trackers are not cheap. The cost of the tracker may be used to purchase a fixed rack and more PV panels, which might offset the loss in non-tracking production.
- Trackers add a degree of additional complexity to the system. The bits and pieces are just one more thing to have to maintain.

Regardless of which type of PV mounting system you decide to use, keep in mind that they take up a fair bit of area and make wonderful kites or sails in high winds. Ensure that proper mounting and foundation work has been undertaken in compliance with the manufacturer's installation instructions. If you want to harness the wind, don't use your PV panels.

Step 4 – Eliminate Tree Shading

Although we want to shade the south side of our house from the summer sun, the last thing you want to do is shade even a very small portion of your PV array. Partial shading will cause a disproportionate reduction in electrical generation to shading area. Keep trees well clipped in the sunlight window, keeping in mind both winter and summer sun tracking.

Step 5 - Consider Snow and Ice Buildup in Winter

If you live in the northeast, consider PV location with respect to winter snow and ice buildup. Although PV angles at this time of year are almost vertical in this area, snow and ice will stick to the array. If a coating of fresh white powdery snow is covering the array it may take several days for it to fall off. PV arrays mounted on the second story roof are not likely to be wiped off.

Step 6- Ensure Modules Do Not Overheat in Summer

PV module electrical output fades (just like everyone I know) as the mercury rises. In order to ensure peak operation of the array, modules must not be seated directly on the roof surface. An air gap of 2 or 3 inches (5 to 8 cm) will ensure cooling air can circulate under the array.

Step 7 – Decide on System Voltage

For off-grid systems, there are three commonly used system voltages: 12 V, 24 V or 48 V. Although it is possible to select differing voltage and power levels, the industry uses the following guidelines.

System Size	PV/Battery Voltage
Up to 2 kWh per day	12 V
2 to 7 kWh per day	24 V
Over 7 kWh per day	48 V

The PV industry has standardized on modules with 12 V nominal outputs. As we discussed earlier and as shown in Figure 6-4, we saw how to interconnect PV modules in parallel to increase current flow and system wattage without changing the voltage of the panels. Figure 6-9 illustrates a mixture of series- and parallel-connected PV panels. In this arrangement, there are two rows of four PV panels. Both rows are interconnected in series, increasing the system voltage. If you look carefully and pretend that each PV module is a battery, you can follow the series connection. The positive of one module connects to the negative of the next, and so on. When each 12 volt panel is connected in a series grouping of four, the voltages are additive, resulting in a 48 volt array.

8 PV modules connected in series and parallel

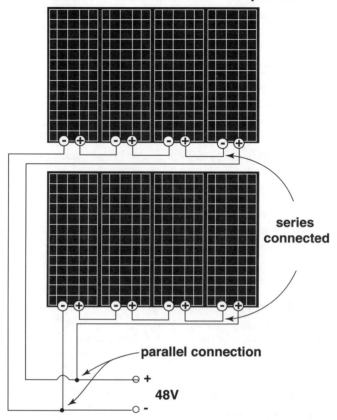

Figure 6-9. PV panels are just like batteries: they can be interconnected in series to increase voltage, and in parallel to increase the current of your array.

Once both rows of panels are at the same system voltage, which in this case is 48 V, the array may be connected in parallel. Examine the wiring connection at the point indicated in Figure 6-9. The negative lead from each row is connected, as is the positive lead. Together, the two rows of 48 V arrays will deliver more flow of current to the load.

Step 8 - Locate the Array as Close to the Batteries or Inverter as Possible

Have you ever connected 2 or 3 garden hoses together, turned on the tap and wondered why the sprinkler was dribbling an anemic spray of water, even though the pressure at the house was fine? What you witnessed is resistance to flow. This phenomenon is not just related to garden hoses but to electrical circuits as well. The resistance to flow of electricity works in the same manner. Low voltage (pressure) circuits require very large wires to carry the necessary current in the cable. It stands to reason that the further the PV array is from the battery or inverter, the greater the loss of electrical energy will be. Increasing the pressure or voltage of the electricity will help somewhat, but distance is a bad word when it comes to low-voltage direct current circuits. Keep the wire runs as short as possible. Total wire length will be discussed in more detail in *Chapter 11 – Electrical Wiring Issues*.

Conclusion

PV systems are simple, reliable and they work! Where a home or cottage is just a little too far from the electrical lines to connect, there is no contest. It will make economic and environmental sense without further consideration. Grid-interconnected systems require a little more work performing economic payback. With electrical rates rising and governments committed to "greening the grid", there is little question that all of North America will be pushing PV in the same manner as California is now.

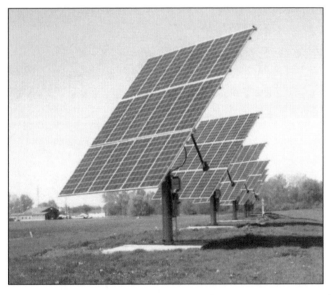

Figure 6-10. The entire renewable energy system is known as "PV on a stick". All of the electrical controls, PV array and sun tracker are interconnected to the grid in one simple arrangement. (Array Technologies)

Sun-spiration

"During the ice storm of 1998, millions of people were left without electricity for extended periods of time. In our rural area, neighbors were left powerless (figuratively and literally) for over 3 weeks.

Because many trees had fallen across laneways as well as damaging phone lines at the same time, there was a serious concern for elderly people who may have been shut in and unable to call for help.

The army was sent door to door to canvas if anyone was in trouble. On a clear Sunday morning an army jeep arrived in our laneway. I quickly finished making a cappuccino and went to the door to greet them.

One officer started walking to the house, with one eye suspiciously watching the spinning wind turbine. He gazed at my freshly made coffee and possibly noted some Mozart floating out the door. With a rye grin he said "I just have to ask..... Is everyone OK here?"

I smiled broadly as I explained that life just couldn't get better".

Bill Kemp

7. Electricity from the Wind

I have been told by a number of people that while women seem to like PV technology, men tend to gravitate towards wind turbines. Perhaps, for the men, it has something to do with the big tools, concrete, propellers and spinning stuff that provides the appeal. Whatever the case, mankind has been capturing the wind for eons. Every school child knows of the fabled Dutch windmills. What they may not know is the Dutch, along with their neighbors from Sweden and Denmark, are among the world's foremost users of wind electricity generators.

North America is catching up. Years of low fossil fuel prices and certainty of supply convinced most North Americans that there was nothing to worry about. The Europeans on the other hand, have been

Figure 7-1. This 10 kW wind turbine is typical of larger home-based machines. With a blade diameter of 23 ft (7 m) and the right amount of wind, this unit can provide sufficient power for almost any home. (Bergey WindPower).

worrying for years. Most European countries rely on imported oil and have high population densities. Conserving electrical energy has become a way of life there.

Fortunately we are waking up on this side of the pond, and as the alarm bells are ringing, wind turbines are coming of age. Sites, such as King Mountain in West Texas, are springing up here and there, and are good examples of us catching up. Like everything in Texas this site is big, having 214 wind turbines installed. Together they provide an output of 278,000,000 W (278 MW), which is sufficient to power 80,000 super-sized Texas homes or to support 3% of Denmark's electricity requirements. These units are not toys.

It is doubtful that you require this much power for your renewable energy powered home, but it does give you some scope of the possibilities of using wind turbines.

Home-Sized Wind Turbines

The market today is alive with many manufacturers of wind turbine systems. There is also a market for rebuilt and used machines that were installed prior to rural homes being connected to the

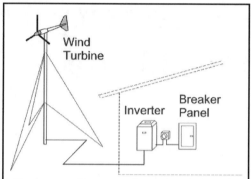

Figure 7-2. Home-sized wind turbines may be connected to the electrical grid in the same manner as small PV systems. (Bergey WindPower)

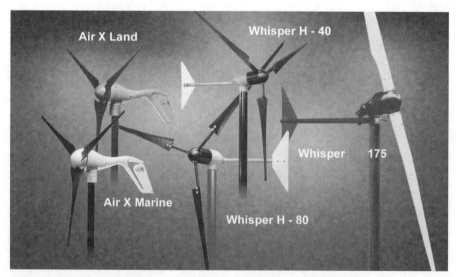

Figure 7-3. Wind turbines are available in a wide variety of sizes, designed to suit just about every application. (Southwest Wind Power)

Figure 7-4. Wind turbines are used extensively in this obvious off-grid installation. What better harmony than using the wind to sail the boat and make electricity at the same time. (Southwest Wind Power)

growing electrical grid in the early 1900s. The Resource Guide in Appendix 3 provides a listing of the major manufacturers and re-builders of new and refurbished equipment.

When people see a wind turbine, their first question is usually, "How big is it?" This is the most misleading question asked when considering what "size" of turbine is required for your application.

Let's start our review of wind turbines by (figuratively) taking a unit apart to understand how they are manufactured and operate. Unlike PV module technology, wind turbines require a greater amount of assessment to fit your application.

The majority of home wind turbines are known as horizontal axis machines, similar to the one shown in Figure 7-5. The major components' names are common amongst all wind turbines, regardless of size. A tower structure supports the wind turbine, raising it high above the land on which it is installed. The higher the tower, the greater the speed and smoothness of the wind, and therefore, the greater the power. Typical tower heights range from 60 to 100 ft (18 to 30 m). The tower can be self-supporting, or more commonly and less expensively, it can have a series of guy wires used to support it. A concrete pad, which bears the downward weight, also supports the tower and prevents it from skidding sideways. The guy wires provide all of the support and, as such, are secured to earth or concrete foundations using anchors. Cable tensioning devices are installed on each guy wire to remove slack and prevent tower motion.

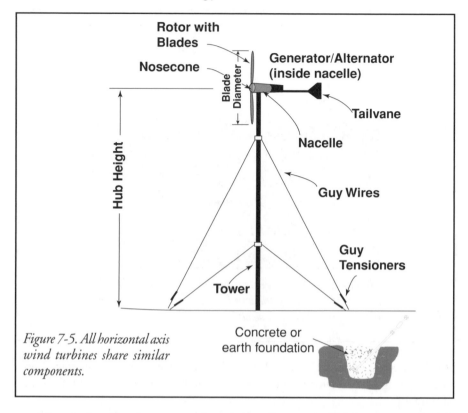

Rotor with
Blades

Nosecone

Generator/Alternator
(inside nacelle)

Blade Diameter

Tailvane

Hub Height

Nacelle

Guy Wires

Guy Tensioners

Tower

Concrete or
earth foundation

Figure 7-5. All horizontal axis wind turbines share similar components.

The wind turbine has some rather obvious features itself. The most noticeable are the rotor and blade assembly. The rotor is the mechanical assembly in the centerline of the turbine. This component holds the blades on the outside and the generator/alternator on the inside. A nosecone is attached to the rotor to provide a smooth surface for the wind to pass over as it travels through the blades. A *nacelle* or cover protects the generator and support components from the weather. Finally, a tailvane assembly is used to keep the rotor and blade

Figure 7-6. The blades of a wind turbine are shaped into a form similar to an aircraft wing. Wind rushing past the blades causes aerodynamic lift that imparts a torque or rotational force powering the generator. (Bergey WindPower)

assembly pointed into the wind. On some models, the tail-vane also provides a means of slowing the blades down in dangerously high winds by automatically performing a function known as *furling*. On models such as the Bergey Excel, shown in Figure 7-1, a manual furling winch is also provided, allowing the unit to be shut down during severe storms.

A close up view of a small 1,000 W turbine assembly is shown in Figure 7-7. From this view, you can clearly see the rotor assembly with the blades removed from their sockets. The generator wiring is just visible at the rear of the rotor assembly. If you follow the wires from the generator, you can see them terminate into a black module, which is known as the *rectifier bridge*. This device converts the alternating current from the generator into direct current suitable for direct connection to the battery bank. The alternating current is input as three-phase (which is similar to most industrial building services) at the top of the rectifier; the direct current output is at the bottom of the rectifier. Following the positive and negative lead wires from the rectifier, you can see that they reach a section of the tower assembly. To prevent the power cables running down to the house from twisting as the turbine spins around on the tower's top section, the cables are fed to a "slip ring" assembly. Slip rings act as brushes, conveying the electrical power from the rotating *yaw tube* to the stationary tower head assembly. The tail-vane completes the assembly and mounts onto the rear post.

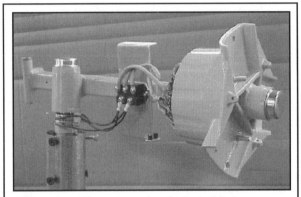

Figure 7-7. This picture details the inside frame, rotor and generator of a small 1 kW wind turbine, and shows the blades and nacelle cover removed. (Bergey WindPower)

Wind Turbine Ratings

Now that you have been formally introduced to a typical home sized turbine, we need to clear up some of the details regarding size. Unlike the debate over certain biological issues, size in a wind turbine *does* count.

It's actually the diameter of the wind turbine blades that truly counts. You may recall from *Chapter 3 – Energy 101* that in order to convert kinetic energy to electrical energy, we have to capture as much of the "moving fluid" as possible. The blades catch the kinetic energy of the wind. A unit with a smaller blade area captures a smaller amount of energy. It's that simple. Because the blades sweep in a circular path, we have to do a little bit of fancy mathematics to calculate the area they sweep. Assume our turbine blades have a diameter of 10 ft (3 m), giving them a radius (half the diameter) of 5 ft (1.5 m).

π x *Radius* ^{2} = *Area Swept by Blades*
3.14 x 25   = *78.5 square ft (7.3 m²)*

Machine manufacturers like to use power ratings that are not as well standardized as with PV modules. This makes comparison shopping difficult. Calculating machines' rotor area sweep and comparing these values provides an even keel for calculating power output. Keep in mind that rotor sweep area has a linear relationship with power. Simply stated, a doubling of the sweep area doubles the power output.

If the math is too much bother, an alternate means of comparing machine ratings is by consulting the "power curve" graph. The graph shown in Figure 7-8 represents the large home sized turbine shown in Figure 7-1. This graph details the electrical power output in kilowatts based on a given wind speed in miles-per-hour (mph). For example, at a wind speed of 24 mph (38.5 kph), this unit's output is 6,000 W of electrical energy. An excellent way of cross-checking manufacturers' figures is to compare both power curves at the same wind speed as well as the rotor area sweep. The ratio between the two should be approximately the same: a ma-

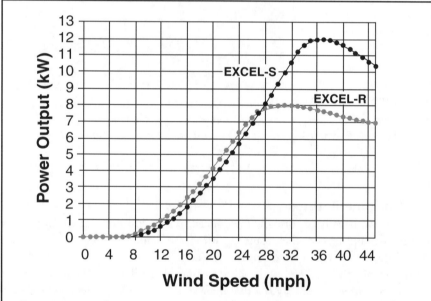

Figure 7-8. An alternate way to compare wind turbine output is by consulting the power curve rating. This chart shows the electrical power output for a given wind speed. (Bergey WindPower)

chine with twice the rotor area should have a power curve with twice the power output.

The power curve graph also details some other interesting facts concerning all wind turbines. You will notice that at low wind speed, below approximately 8 mph (13 kph) the turbine output is almost zero. Many dealers will talk about units that start in low-wind conditions or begin generating power earlier than others. This information is not factually supported by the physics of wind energy. The turbine has nothing to do with "starting to kick in earlier". Sure the unit might spin a bit in low-wind periods, but it will not output any energy. Energy is related to wind speed and you need a fairly high wind to make energy.

This brings up another important fact about wind energy. The relationship between wind speed and power is not linear. Removing the "tech talk", doubling the wind speed does not double the power. Take another look at the power curve in Figure 7-8. Follow the Excel-S data line at a wind speed of 16 mph (25.5 kph) and we derive an output of approximately 2,200 W of power. Now if we double the wind speed to 32 mph (51 kph) we do not get 4,400 W as you might expect. Check the curve and you'll find that the output is now up to a whopping 10,500 W!

What happened? This is a classic example of the non-linear relationship between wind speed and energy in the wind. In fact, the theoretical relationship is cubic. This means that if the wind speed is doubled, the power output is increased to eight times ($2^3 = 8$)

Rating Jargon

The power curve also details information about how the unit will behave at various wind speeds. Refer to the Bergey model Excel-S power curve in Figure 7-8, and you will see a shape to the curve that will be similar, but not identical, to many manufacturers' units. These points on the curve are discussed below.

- **Start-up speed**. This is the wind speed at which the rotor and blade assembly first begins to rotate. The start-up speed is irrelevant, as this low rotational speed provides no useable power output.
- **Cut-in speed**. Relates to the wind speed that the generator actually begins to produce **useable power**. This is one of the most misunderstood terms in wind power generation. There is no power in slow wind; if the unit starts at 3 mph (5 kph), it will not be generating useful levels of power. Likewise, if the turbine's cut-in speed is being touted as exceptionally low and better than every other model, hang on to your wallet.
- **Rated speed/power point**. The manufacturers' data states the nominal power rating of the unit at a given wind speed. The Bergey Excel-S is rated at 10,000 W and that occurs at a wind speed of approximately 31 mph (50 kph). In practice, the rated speed/power point is of little value as the machine will likely spend very little of its life at this wind level. Additionally, each manufacturer rates their turbines at different wind speeds, skewing the data values and mak-

ing cross comparison difficult. Remember, the rotor diameter/sweep area is the most important rating to be concerned with.

- **Peak or Maximum Power.** This indicates the maximum power the unit will generate. The Excel-S is capable of producing 12,000 W at 36 mph (58 kph) wind speed.

- **Furling speed.** The furling speed is the wind speed at which the unit enters a "self-protection" state. It is characterized by increasing wind speed causing a lowering of output power. In most small wind turbines, the tail and nacelle "hinge" so that the blade/rotor section turns out of the wind. The rotor pointing upwards, similar to a helicopter, until wind speed subsides may also accomplish furling.

Ratings are important, but the most important consideration is to determine what point on the power curve the turbine will work at *your* site. Don't become overly confused by the extra jargon that will become immediately unimportant the minute you switch the unit on.

Wind Resources in Your Area

The wind tends to vary greatly over the span of a year and depends largely on where you live. There is almost no wind in central Florida, but it never stops blowing in Cape Cod. If a small change in wind speed causes a large change in electrical power output, then we must be careful to capture the highest wind speed. Almost everyone I have talked to about installing a wind turbine tells me that they are on a hill or near a lake and have "great wind". But just what do they mean by "great wind"? A 9 mph (14.5kph) breeze is pretty strong. At this speed, it will make holding your umbrella a bit difficult or whip your tie around your face, so it must be a good wind turbine site. Think again. Looking at the power curve for the 10,000-Watt Bergey Excel model in Figure 7-8, you will find the power output is nearly zero.

You can't take chances installing a turbine and hoping it will work out. This is the surest way to electrical and possibly financial bankruptcy if there ever was one. Having someone tell you there is plenty of wind is just not going to work. A wind turbine is a big investment and a careful site study should be conducted before you commit your dollars and time.

There are two typical methods of determining the characteristics of the wind at a given site. The first method relies on wind maps prepared for the government; the second entails a site measurement study over a period of a number of months or longer. Before we discuss wind mapping and surveys, we first need to review what is meant by wind assessment.

A wind turbine without wind is like the proverbial fish out of water. Wind makes a wind turbine operate and, within reason, the more the better. But what wind are we talking about? Wind in the spring is always strong, while in the middle of summer it can be stifling without a breeze to be found. The answer is related to need. If you have a hybrid system with a PV array, it is quite likely that you will require no additional energy during the "bright" late spring, summer and early fall period. Late fall and winter tend to be dark and rather tough on a PV system, so adding a bit of wind energy to boost things up would be helpful then. On the other hand, a system that relies on wind as the main or only source of energy will require a steady supply all year round, with an emphasis on fall and winter when more time is spent indoors and energy needs tend to be higher.

Manufacturers publish tables that estimate the monthly or yearly energy production in kilowatt-hours of energy when the turbine is installed on a tower of given height and at differing wind speeds. An example taken from the Bergey Excel-S is shown in Table 7-1.

Tower Height	Average Wind Speed						
	8mph (13 kph)	9mph (14 kph)	10mph (16 kph)	11mph (18 kph)	12mph (19 kph)	13mph (21 kph)	14mph (23 kph)
60 ft (18 m)	330 kWh	480 kWh	670 kWh	870 kWh	1,110 kWh	1,350 kWh	1,610 kWh
80 ft (24 m)	430	620	840	1,100	1,370	1,670	1,960
100 ft (30 m)	490	700	950	1,220	1,510	1,820	2,130
120 ft (37 m)	550	780	1,050	1,340	1,650	1,970	2,280

Table 7-1. This chart details the relationship between wind speed, tower height and monthly energy production (in kWh) for the Bergey Excel-S turbine. (Bergey WindPower)

As the wind speed increases, so does the amount of energy produced. But what does the height of the tower have to do with energy production? Tower height relates indirectly to wind speed. Think of wind as water for a moment. If water were streaming over your wind turbine and it were close to the ground, trees, rocks, buildings and other obstructions would cause the water to create rapids and generally slow down. As you move above the "river bed" example, there are fewer obstructions and the water would move much more smoothly. At some point, there would be no obstructions in the water's path and the flow would become very smooth. At this point we have what is called *laminar flow*. Laminar flow not only increases wind speed and energy production, it also prevents "rough air" from stressing the turbine, ensuring longer life.

A general rule of thumb is to mount the turbine on a tower 30 ft (9 m) above any obstructions that are within a 300 ft (90 m) radius of the tower. For example, the house and trees shown in Figure 7-9, which are within a 300 ft radius of the tower, must be at least 30 ft (90 m) shorter than the turbine on the tower. Keep in mind that trees grow, so provide enough margin of error in tower height or use

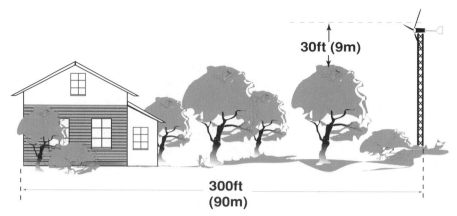

Figure 7-9. A wind turbine must be placed in smooth air, clear of any obstructions.

pruning equipment to maintain the height of the trees.

The effect of tower height on turbine power cannot be under-emphasized. At 8 mph (13 kph) wind speed, the Bergey Excel-S power output increases from 330 kWh to 550 kWh moving from a 60 to 120 ft tower. In other words, an increase of tower height from 60 to 120 ft increases power output by 1.7 times.

Before you rush out and purchase a 700 ft (213 m) super-tower, bear in mind the tower cost. Foundation work and wiring expenses, etc. will tend to throw cold water on your plans. A tower of 60 ft to 80 ft is a good start for smaller units under 1,000 W. Up to 120 ft is recommended for towers with larger turbines, but work within your budget.

Wind Mapping

The power output chart shown in Table 7-1 lists energy output at varying wind speeds. The wind speed references are based on yearly average wind levels measured at a standard height of 33 ft (10 m) using a device known as an anemometer, such as the one shown in Figure 7-10. This device consists of a series of cups that are arranged so that greater wind speed causes the cups to spin faster. The spinning shaft is connected to a speed detection device, the output of which is fed to a data logger or recording instrument, recording wind speed. The wind speed data is collected for long periods of time and is then plotted using a mathematical *Rayleigh distribution* curve. The curve plots recorded wind speed at the frequency of the occurrence. The highest point on the curve is the speed at which the wind blows the most often. When this data is recorded over a 12-month period, the result is the average annual wind speed.

Maps are available which detail wind speed annually, seasonally and monthly. These maps are compiled by and for the government for many diverse needs such as crop planting, weather forecasting and wind turbine installation. A typical wind energy map is shown in Figure 7-11, which details mean annual wind resources in

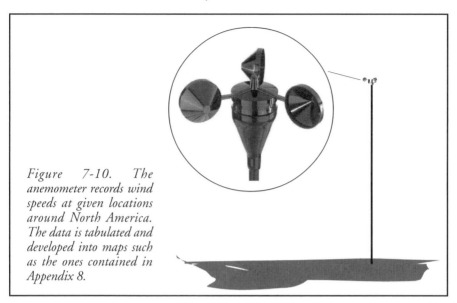

Figure 7-10. The anemometer records wind speeds at given locations around North America. The data is tabulated and developed into maps such as the ones contained in Appendix 8.

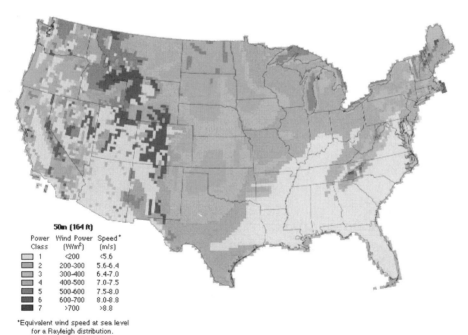

| 50m (164 ft) | | |
Power Class	Wind Power (W/m²)	Speed* (m/s)
1	<200	<5.6
2	200-300	5.6-6.4
3	300-400	6.4-7.0
4	400-500	7.0-7.5
5	500-600	7.5-8.0
6	600-700	8.0-8.8
7	>700	>8.8

*Equivalent wind speed at sea level for a Rayleigh distribution.

Wind Energy Resource Atlas of the United States

Figure 7-11. This wind map of the United States shows annual average wind speed. Each area is broken down into different classes. A minimum power class for off-grid homes is Class 2. (Courtesy Bergey Windpower)

the United States. A full map for North America is shown in Appendix 8. Wind speed maps have been broken down into areas of speed and "power class". The typical rating of speed is in meters-per-second, which can be converted to other units:

Wind Speed meters/second	Kilometers per hour	Miles per hour	Quality of Site for Wind Power	Power Class
4	14.4	8.9	Not Acceptable	1
5	18	11.2	Poor	1
6	21.6	13.4	Moderate	2
7	25.2	15.7	Good	3
7.5	27	16.8	Very Good	4
8	28.8	17.9	Excellent	5
8.5	30.6	19	Excellent	6
9.0	32.4	20.1	Excellent- HI	7

Table 7-2. Wind speed is usually provided in meters-per-second. Use this chart to convert to speeds in miles- and kilometers-per-hour.

Power class numbers that range from one to seven indicate relative energies in the wind based on annual average wind speed for that area. Power Class 1 is considered unacceptable for wind turbine installations. Local geographic considerations such as large bodies of water, hilly or mountainous areas and tree cover can change the average data up or down from the values indicated. Another consideration in determining accuracy of the mapping is the distance from the recording site to your location. Wind mapping is a reasonable method of determining relative wind speed, but it is not foolproof.

Site Measurement Study

Perhaps the best way to assess your site is to take your own site survey. This is recommended where a larger turbine will be installed for profit, or if the wind mapping data for your location is questionable. Companies such as NRG Systems (www.nrgsystems.com) offer their Wind Explorer model on a 10-meter tower kit for approximately $1,200 (U.S.). These units are battery powered and contain a data logger that will record wind patterns and provide a direct output of wind resources. For even more accurate assessments, larger tower kits are available to place the anemometer at exactly the correct tower height.

Once you have determined what the average wind speed is for your area, you can apply this information to the turbine's power curve. Using the Bergey Excel-S as our model turbine, let's work through a simple energy calculation.

Assume that your site turns out to be a Power Class 2, with annual average wind speeds of 6 meters-per-second. Consult Table 2 and you'll find that this wind speed translates into 13.4 mph (21.6 kph). Locate 13.4 mph on the wind speed or "x" axis of graph 7 - 8 and see that it correlates to a power output of approximately 1,000 W. We can now roughly estimate the expected yearly energy output (energy = power x time):

a) 1,000 W output x 24 hr/day x 365 days/year = 8,760,000 W hours/year

b) 8,760,000 W hours/year ÷ 12 months/year = 730,000 W hours/month

c) 730,000 W hours/month ÷ 30 days/month = 24,000 W hours/day

The Bergey energy chart in table 7-1 shows a higher value. Comparing average wind speed with instantaneous power output of the turbine **does not** take into account the affects of higher and lower windspeed from the average. This simplified approach will always underestimate energy production. Possibly the most interesting item relates to the estimated power output per day. A large renewable energy system consumption is 10 kWh per day. The Bergey Excel-S is providing more than twice this amount of energy, even when it is located in a "moderate" wind region. Based on these calculations, take a look at a smaller, less expensive turbine. You may have to run through this calculation or compare estimated energy output charts a few times for different models, but the exercise will pay off by ensuring that you purchase a machine that is just the right size.

Site Location and Installation

Another major difference between PV and wind-based systems are their size and height. You can a hang a few PV panels on almost any roof (with "few" being the keyword); wind turbines on the other hand, will attract attention for miles around. If those "miles around" happen to be forests or grazing lands, then only the cattle will pay attention. But don't even consider installing a wind turbine until you have all the "Is dotted and Ts crossed". Not everyone will share your enthusiasm for wind and some may just simply object out of fear.

Where you locate the wind turbine will generally be determined by your lot size, cable distance to the main building and tower construction. The best location however is where the winds blow strongest, or wherever your pocket book says to put it. Wind speed and smoothness (or laminar flow) will be better over unobstructed grazing land in contrast to forested areas, owing to ground surface smoothness. Tall trees in a wood lot or forest increase the "working" ground level to the tree height, which in effect reduces the tower height. Follow the general site requirements in Figure 7-9 to ensure proper placement.

If the tower is free-standing, such as the one shown in Figure 7-1, it can be installed almost any place that will accommodate the concrete base. More often than not, the tower will be a tubular guyed design or tripod lattice tower, as shown in Figure 7-12. Guyed towers are inexpensive and tend to be the most popular.

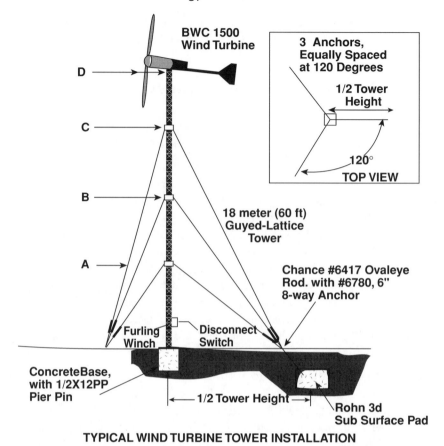

TYPICAL WIND TURBINE TOWER INSTALLATION

Figure 7-12. Guyed lattice and tubular towers are commonly used for small wind turbine installations. Their main drawback is the room required for guy wire clearance.

The guy wires, however, necessitate considerably more room for installation. A good rule of thumb is to allow a minimum clearance radius of half the tower height for guy wire placement.

A power feed cable is required to connect the wind turbine to the battery bank or inverter. An aerial cable may be run between the tower and house. More typically, a direct burial cable or cable in a PVC plastic conduit is used. The selection is based simply on preference and obstacles in the cable path. It may be simpler to provide an overhead cable than to blast rock or dig up a paved laneway. Regardless of which method is chosen, there are specific electrical code requirements necessary to ensure fire and shock safety. These items are discussed in more detail in Chapter 13.

Zoning and building permits may be required depending upon your proximity to urban areas. Wind turbines are almost exclusively installed in rural areas, where zoning restrictions are minimal at best and building permits are relatively easy to obtain. Regardless of whether the installation is rural or urban, the best advice is to plan things properly from the start.

A neatly drawn site plan showing property lines and setback distances is a good starting point. Building officials will almost certainly not know anything about wind turbines. This may help or hinder your cause. It may help because they are curious and willing to assist with something "neat". On the other hand, their lack of knowledge may also cause problems, no matter how silly. Don't be surprised if you are asked if a strong wind might blow the tower down, or if aircraft will become tangled in the unit. It pays to do a bit of homework.

- Provide some pictures of local radio/TV antennas that are of similar or greater height. A wind turbine tower may seem odd, yet most people can't even remember seeing a 200 ft tall microwave tower just down the street.

- Ask your tower manufacturer to provide pre-engineered drawings of suitable installation plans. Your building inspector is used to walls and roof trusses, not towers. Pre-engineered drawings will provide a level of comfort that the unit isn't going to crash through your roof during the next windstorm.

- If you have neighbors, be sure to get them onside first. A wind turbine shows a level of stewardship and concern for the environment that should be welcomed. But if George next door thinks it will kill all the birds, interfere with his TV signals and reruns of Baywatch, you may be in for some opposition. These concerns are unfounded at best. Another common concern is the noise produced by the units. This is not a problem; by the time a wind turbine is really working (and yes, it does make *some* noise), the trees, windows and houses are likely creaking at a far higher volume level.

- Check for legal easements or right-of-ways deeded to your property. This may include underground gas or telephone cables and the like. If you are unsure, check with your lawyer and show him or her your site plan. No one wants to pay a legal bill, but it is a lot less expensive than moving your tower over a couple of feet.

- Height restrictions may come into play if you are in within a certain distance of an airport or if there are special zoning ordinances. Zoning variations should not be a problem, especially if there are radio towers or buildings in the township that are of similar height. Although aircraft interference is a minor issue, it is a good idea to check with your local FAA office in the United States or NavCan in Canada. Paint and lighting restrictions may be enforced in some areas, but usually only when the tower is over 200 ft (61 m) high in the United States and 80 ft (24 m) high in Canada. Both agencies will provide a letter indicating whether or not markings will be required.

You have every legal and ethical right to install a wind turbine, in the same manner as planting a tall tree or erecting a TV antenna. Problems with neighbors and building officials occur because of ignorance or a lack of planning. Since you are enthusiastic and better informed, take the time to transfer this knowledge and educate those around you. The results should lead to an amicable agreement between the concerned parties.

Getting Started and a Word About Safety

Once the wind turbine, tower and site location have been selected, the next step is to start planning the installation. Erecting even a small wind turbine is a job that can be dangerous and tax the skills of the most mechanically inclined. But don't let this dissuade you; completing the work yourself is satisfying and will save a considerable amount of installation cost.

There are three common methods of erecting a wind turbine and tower. The first uses a tilt-up arrangement that allows the tower to lie flat to the ground during installation and servicing. When the tower is ready to be erected, a hand or electric winch can be used to pull the tower to a vertical position. This system is primarily used with smaller wind turbines (<1 kW) and shorter tower sections. Small winch-up towers are guyed, while larger "tubular" towers are becoming available which have elegant tapered poles and can be hinged to allow lowering for service.

The second type of installation involves building the tower on the ground, and then using a crane to erect the tower in one piece. The advantage of this system is no one has to leave the ground and work up high on the tower. The disadvantage is that servicing requires taking a (second) mortgage out on the house to pay for the crane.

The third method requires the use of a *gin pole* and some tower climbing theatrics that would rival the gymnasts of Cirque du Soleil. A gin pole is a boom or pipe that is bolted to the top of an installed tower section that extends upwards to where the next section is to be placed. A ground crew using a pulley mounted to the top off the gin pole raises the next tower section, while someone who is strapped to the tower gingerly connects the successive sections. Notwithstanding my personal bias, I believe the name *gin pole* reflects the need to imbibe such a libation as gin after successfully completing this death-defying act.

I cannot emphasize enough how difficult working on a tower can be. Climbing a long extension ladder is nothing compared to climbing and working on a tower. A ladder always has some horizontal slant, making this more akin to walking up steep stairs. Safely climbing a tower with work boots, fall-protection gear and a bag of tools is great provided you spend your weekends rappelling El Capitán, but it can be very tiring and dangerous work. I would stick with the tilt tower or mortgage the house and hire a crane before recommending a gin pole system to anyone.

At some point it may be necessary to actually climb the tower. Before you get to the top and realize you are too scared to let go and actually do some work, it would be wise to discuss climbing gear.

There are too many people who think safety equipment isn't cool or even necessary. The smallest wind turbine tower for home use will be upwards of 60 ft (18

m) high. That's about the same as falling from a 5-storey balcony; it is unlikely that you would survive. Visit your local safety supply store where for a few dollars you can purchase an approved full body/ climbing harness. The Occupational Safety and Health Administration in the United States and Canada both certify climbing gear suited to this work. Do not purchase a "safety belt" or other non-approved device. One slip off the tower rung and you may be dumped upside down and slide right out of your safety belt.

A safety harness is of no value if it isn't connected to something, preferably the tower. A short section (< 6 ft / 2 m) of nylon rope may be

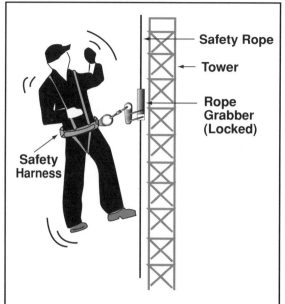

Figure 7-13. Falling from a 100 ft wind turbine tower is not something most people find enjoyable. A safety harness and rope grabber, such as this model shown here, makes climbing and servicing a much safer task.

attached to the harness and then successively clipped and unclipped to the tower using a snap hook as you climb. While Steel snap hooks and approved nylon rope or lanyard will work, there are drawbacks to this system.

The first drawback relates to the fact that constantly clipping and unclipping the hook gets in the way, and there is a tendency to not use the clip until reaching the tower top. Climbing the small round legs of a tower can be difficult and slippery. During the descent, you may be quite tired after spending time at the top. Safety is just as important on the climb downward as it is on the climb upward.

The second problem is with the length of lanyard required between the tower and safety harness. A lanyard that is 6 ft (2 m) long will allow you to drop that same distance before breaking a fall. This may be enough of a drop to cause serious back or muscle injury, making recovery difficult. Likewise, a lanyard of steel wire will not give, causing an abrupt stop causing further injury.

The best value for your money and life is a device from Flexible Lifeline Systems (www.fall-arrest.com). These units comprise a long nylon rope or lifeline, which is attached to the top and bottom of the tower. A "rope grabber" is placed on the rope and in turn clipped to the full body harness. When you climb up the tower, the rope grabber slides along allowing for easy ascent.

Should you fall, the weight of your body causes a cam inside the rope grabber to pinch the lifeline breaking your fall, as shown in Figure 7-13. During a normal descent, the rope grabber must be lifted and slid down the lifeline one section at a time throughout the descent.

This system will limit your freedom of motion at the top of the tower, due to the lifeline being affixed to the tower top section. To provide a bit more freedom, a short lanyard may be attached between the rope grabber and the safety harness.

In addition to protecting yourself from a fall, there are a few other points of safety to be considered.

- Always carry tools in a fitted and sealed pouch. You will require free hands while climbing the tower structure.
- Wear leather gloves and non-slip work boots. Tower rungs are very rough and can be slippery from dew, leaves, pollen and dust.
- When climbing the tower, only one hand or foot should be off the tower at a time. For example, release a hand and reach for the next highest tower rung. Make sure you are gripping the rung before lifting a foot to climb to the next rung.
- Always immobilize the turbine by applying the brake switch and/or furling the machine. Never work on a windy day.
- Wear a hard hat. A light breeze can easily cause the turbine to pitch around or *yaw* striking you in the head.
- Never work on the tower alone. Make sure a buddy is on the ground to assist when needed.
- Make sure that the ground crew is not standing in the way of errant nuts and bolts. A bolt dropped from 100 ft (30 m) is not much different from a bullet being shot from a gun.

Being cheap or stupid when is comes to safety is a no-win proposition.

Tower Foundations and Anchors

Self-Supporting Towers

The self-supporting tower shown in Figure 7-1 is erected on a single pad of concrete reinforced with metal rebar. Alternatively, three holes may be drilled and filled with a rebar and concrete mix providing the necessary pads. It is not possible to fully cover the design of freestanding tower foundations in this Handbook, as there are too many variables to consider. These issues include tower height, weight, wind load, soil type and so forth.

The turbine or tower manufacturer is often able to provide foundation plans that have been pre-approved for most local soil conditions.

Guy Wire Supported Tower

Guy wire supported towers, such as the unit shown in Figure 7-12, are the most common design owing to their lower cost. Guyed towers can be provided in tilt-up

Figures 7-14 a, b, c. The successive steps in pouring a center pier to hold a guy wire supported tower include a) forming the pier excavation, b) pouring the concrete, and c) aligning the center pin and smoothing the surface. (Bergey Windpower)

designs or installed with a crane or gin pole arrangement. Smaller turbines, typically less than 8 ft (2.5 m) diameter, can be mounted on a pipe tower. These towers use very inexpensive 2.5-inch o.d. steel pipe. The turbine manufacturer supplies the guy wire and hardware kit, you purchase the pipe locally.

Lattice towers, such as models manufactured by Rohn Industries (www.rohnnet.com), are more costly than pipe towers, but offer the advantage of fewer guy wires and the ability to climb the tower when servicing is required.

The tower mast sits on a compressive foundation known as a pier. The function of the center pier is to prevent the tower from pressing itself into the ground and from slipping sideways. It offers no vertical support to the tower.

The guy wires are connected to anchors that are spaced 120 degrees apart from each other, as noted in the top view of Figure 7-12. Pipe towers, being structurally weaker, generally require a spacing of 90 degrees. The anchor is spaced a minimum of half the tower height away from the center pier.

The pier is most often formed using concrete, although it is possible to drill a hole in exposed bedrock and mortar a tower support pin into the hole. It is not necessary to create forms to hold concrete, provided the soil will not cave in when pouring begins. Figure 7-14a shows a simple hole dug into the ground, using 2 x 4 inch lumber to create a "finished" look to the surface of the concrete. The hole is lined with rebar rod which helps prevent the concrete from cracking. Rebar is sold in lengths that can be cut with a hacksaw and held together with wire to form a metal "cube". The rebar cube should be supported so that it is completely immersed within the concrete.

Provided your excavation is less than approximately 1 cubic yard or less, it is possible to hand mix the concrete, although most prefer to have a ready mix truck deliver and pour the pier and anchors in one shot, as shown in Figure 7-14b.

A metal rebar pin is shown protruding from the center of the pier in Figure 7-14c. The pin is cut just above the surface of the concrete once it has set, and supports the base of the tower. The sole purpose of the pin is to prevent the tower from sliding off the pier.

Anchors

Smaller wind turbines of less than 60 inches (1.5 m) in diameter may be anchored using "screw-in-place" augers. As the wind turbine size increases, the augers must be encased in concrete, as shown in Figure 7-15. The A.B. Chance Company (www.abchance.com) is a major supplier of anchors to the electrical utility market. They provide an engineering support service on their website which discusses soil types and matching anchor designs. Screw-in augers may be used in dense clay, sand and gravel and hard silts. When in doubt, stick with concrete. A hole dug 4 ft (1.2 m) deep (or below frost line) by 2.5 ft (0.8 m) in diameter must be filled with 2 ft (0.6 m) of concrete to cover the auger helix and will provide sufficient holding strength. It is important to ensure the auger is fixed below the frost level of the local area, otherwise the anchor may be subjected to "jacking". Jacking is the tendency of buried objects to be pushed towards the surface during the freeze-thaw cycle. Ensure the concrete hole is lined with rebar rod to support the concrete.

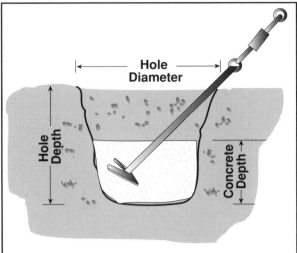

The auger must exit the ground at an angle of approximately 45 degrees and point directly towards the center of the tower. This ensures that the bending force placed on the auger by the guy wires is equal. If the angle is too great or too small, the auger will have a tendency to bend up or down, reducing its strength over time.

Figure 7-15. A guy support auger may be either screwed directly into suitable soil or cemented in place where soil is loose.

Guy Wires

Guy wires may be supplied as part of the tower "kit" or they can be purchased in bulk rolls for custom fitting. Use Extra-High-Strength steel cable for supporting the tower. Refer to the tower manufacturers' data sheets for the diameter of guy cable required for your tower and turbine size. A very conservative size is 5/16 inch (8 mm) EHS steel cable for units up to 3 kW output on a 100 ft (30 m) tower.

The length of guy wire required for a tower depends on the tower height and the radius to where the anchor will be located. Table 7-3 provides the location of the guy wires for each of the levels (A, B, C and D) noted on the installation in Figure 7-12.

The guy wire length required to run from the various heights to the anchor can be calculated using a bit of grade 9 trigonometry called the Pythagorean theorem. As my classmates and I were in a deep slumber during this lesson, we should take a moment to review the calculation.

Flip back to Figure 7-12 and you will see that the tower and finished grade form two sides of a triangle, with the angle at the grade/tower intersection point being 90 degrees. The right angle formed by the two sides and the hypotenuse created by the guy wire form a right angle triangle. Assume that we want to find the length of the guy wire at position "B" for a 60 ft (18 m) tower. Refer to Table 7-3 and find that a 60 ft tower requires a guy wire at position "B" to be 55 ft (17 m)

Tower Height ft / (m)	Height to "A"	Height to "B"	Height to "C"	Height to "D"
40 / 12	17 / 5	35 / 11		
60 / 18	27 / 8	55 / 17		
70 / 21	22 / 7	44 / 14	65 / 20	
80 / 24	25 / 8	50 / 16	75 / 23	
90 / 27	28 / 9	57 / 15	85 / 26	
100 / 30	24 / 7	48 / 14	72 / 22	95 / 29
110 / 34	27 / 9	53 / 16	79 / 24	105 / 32
120 / 37	29 / 9	57 / 17	86 / 26	115 / 35

Table 7-3. The number and height of each guy wire is determined by the overall tower height.

above grade. The guy radius is half the tower height, which is 30 ft (9 m). From these data we can determine:

(Tower Guy height ^{2} + Guy Radius ^{2}) = Guy Wire Length ^{2}

Therefore:

$\sqrt{}$ *(Tower Guy Height ^{2} + Guy Radius ^{2}) = Guy Wire Length*

= $\sqrt{}$ *(55 ^{2} + 30 ^{2})*

= $\sqrt{}$ *(3,025 + 900)*

= *62.5 ft (19 m)*

This length provides the exact distance between the tower and the anchor. An allowance of an additional 5 ft (1.5m) should be provided to allow for misalignment and affixing of the guy to the tower and anchor brackets.

Once the guy wire lengths are determined, carefully unroll the guy cable in a straight line along the ground. Even the slightest kinking will render the guy wire useless, so use caution. Speaking of caution, be sure to wear proper eye protection so that the cable does not whip around and hit you.

The guy wires may be attached to the tower and anchor using U-bolts or pre-formed cable grips. Either method

Figure 7-16. Guy wires for larger machines are connected to a turnbuckle, which allows the cables to be tightened as required. A safety chain or piece of guy wire is threaded through the turnbuckle to prevent unauthorized loosening.

is adequate, although most guy cable suppliers will recommend cable grips due to their ability to be easily adjusted.

Larger wind turbines require the guy wires to be tightened by using turnbuckles, as shown in Figure 7-16. After tightening the cables, thread a safety chain or piece of guy wire cable through the turnbuckle as illustrated to prevent inadvertent or malicious loosening.

Smaller units of less than 60 inches (1.5 m) in diameter do not require turnbuckles. These guy wires may be tightened by using a cable puller or "come-along", as shown in Figure 7-17, prior to applying the guy wire clamps.

Figure 7-17. Guy wires can be tightened initially using a "come-along" or cable puller. Smaller machines of less than 60-inches (1.5 m) in diameter do not require turnbuckles. (Bergey Windpower)

Electrical Supply Leads

Small turbines' output direct current at 12 V, 24 V or 48 V, while large machines output three-phase alternating current, which is rectified at a remotely located "controller box". In order to transfer this power down the tower, the turbine will have a connection terminal or be supplied with a short section of "flying leads". A direct current machine will be equipped with 2 wires, one positive and the other negative. A machine that outputs three-phase alternating current will have three wires, without any polarity or connection concerns. A ground wire may also be provided.

The connections between the electrical leads from the wind turbine and the tower wires should be made in a weatherproof junction box mounted near the top of the tower. The tower to house lead wires should be weather resistant flexible armored

Figure 7-18. The tower and turbine are ready for lifting by the crane. Note the tower bottom plate center hole and mating pier pin at lower left. (Bergey WindPower)

cable (Teck is one example of trade name cable) or be enclosed in a conduit. The power lead wires or conduit may be strapped to one of the tower legs using either ultra-violet protected tie-wraps or cable clamps. Wire size and selection are discussed in Chapter 13.

Erection of the Tower with a Crane

Pre-Lift Check List

At this point, the tower should be lying flat on the ground, or slightly raised and supported along its length if the turbine is already installed. We are now ready to prepare for final inspection prior to lifting.

- Ensure that the guy wires (if used) are attached to the tower and are secure. Guys should be run down to the tower in the direction of the anchors. Ensure that the cables are not twisted or mixed up and are in order from lowest to highest.
- Turbine is wired into junction box at top of tower. Connections are made with tower lead wires.
- Tower lead wire is secured to tower and is sufficiently loose at the bottom to allow for tower movement when aligning the base plate with the pier pin.
- Turbine is securely mounted to the tower and bolt torques tested.
- Tower bolts are checked for security and torque.
- Turbine brake and furling are enabled.
- Turbine is tied with rope to prevent yawing and hitting crane or lift cables.
- Check, check and re-check that all hardware and tools are ready and close at hand. The crane service charges by the hour, even when you have to run to town to pickup a couple of last minute 10-cent washers.
- Check the weather for a still, clear day and call for the crane.

The Lift

Once the crane is ready, the tower lifting straps are threaded through the tower rungs at 75% of the tower height. The tower is then slowly lifted making sure that guy wires, the power feed cable and turbine assembly do not tangle in or hit the crane structure.

If the turbine was not installed at ground level, lift the tower to a comfortable working height and install the unit. It is recommended that this step be completed prior to the arrival of the crane, as this time is billed at rates similar to a New York lawyer.

Once the tower is located over the center pier, the guy wires may be secured to their respective anchors, as described above and detailed in Figures 7-15 and 7-17. It is only necessary to approximately level the tower and tighten the cables at this time. Once the tower is secure, a person is required to climb the tower and remove the crane lifting straps and untie the anti-yawing ropes securing the turbine. Use

Figure 7-19. The crane lifting straps are installed at 75% of tower height and wrapped through the rungs to prevent sliding up the tower. Note the support cradle used to hold the tower off the ground, allowing turbine installation before arrival of the crane. (Bergey Windpower)

caution when removing the anti-yawing ropes to prevent the turbine from yawing and striking you.

The crane may now be packed up, checks written and sent on its way.

Hinged Tower Lifting

There are numerous designs of hinged towers on the market today, including guyed and tubular, free-standing models. The advantages of a hinged tower are obvious, allowing all of the installation and service work to be done safely from the ground. In addition, a crane is not required to perform the lift.

Hinged towers utilize a hand operated or power winch that acts on a levered section of the tower. Refer to specific tower installation instructions.

Final Assembly

After raising the tower, or when the crane has left (or stopped billing), align the tower vertically by using a transit sighting scope as detailed in Figure 7-22. A transit may be rented from a local tool company. These units are often present where construction workers are building roads or checking various angles. A basic unit comprises a tripod stand with an optical sighting scope similar to those used on rifles. The tripod is placed between one and two tower heights away from the center pier and leveled using the integrated water bubble levels and adjustable legs. The vertical cross hair in the scope is then aligned with one tower leg, close to the ground.

The scope is then moved vertically up and down, ensuring the tower leg and cross hair remain parallel or aligned along the entire length.

Guy wires may be adjusted tighter or looser using the turnbuckles to compensate for error. The transit is then moved 90 degrees and checked again from this alternate position. It may be necessary to move back and forth a few times to get the alignment perfect.

If you do not have access to a transit, it is possible to use a standard carpenter's level and check for tower alignment at the bottom section. You can then sight up the tower, placing your eye as close to the tower leg or pipe as possible. Bends in the structure will be obvious and can be adjusted by aligning the guy wires as described.

Once the tower is aligned, tighten the guy wires until there is just a little slack in the cables. A good shortcut for testing cable tension is to strike the cable with a hammer. The resulting sound should be similar to that of a plucked guitar string, and a "wave" should be seen running quickly up and then back down the wire. Don't forget to attach the turnbuckle security chain as outlined in Figure 7-16.

Figure 7-20. The tower and turbine are slowly lifted into position with care so as to not tangle the guy wires or cause the turbine body or blades to strike the crane.

Figure 7-21. Once the tower is secured in position, the lifting straps and anti-yawing ropes are removed. This is also a good time to appreciate the view and congratulate yourself on a job well done.

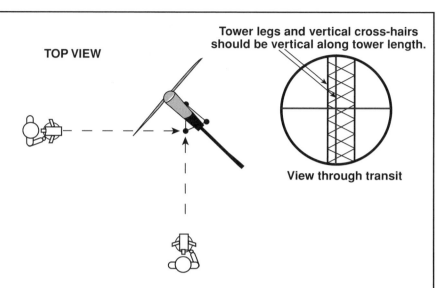

Figure 7-22. Renting a transit is an excellent way to determine if the tower is properly leveled.

Figure 7-23. At least one tower post (more so in sandy conditions) should be connected to a grounding rod driven 8 ft (2.5 m) into the soil.

Figure 7-24. Each set of guy wires should be grounded using a set of U-shaped galvanized steel clamps and attached to the guy anchors. Use smooth bends and as short a run as possible.

The last step in the installation process is grounding of the tower and guy wires. In clay and other moisture retaining soils, a single 8 ft (2.5 m) electrical grounding rod should be driven into the ground as close to or through the tower pier, as is shown in Figure 7-23. In sandy soils, it is recommended that at least two grounding rods be installed at different tower legs to ensure proper lightning protection capabilities.

Each of the guy wires should be connected together using a minimum number #0 gauge grounding wire and galvanized U-shaped bolts, as detailed in Figure 7-24. The end of the ground wire may be connected to the guy anchor or a separate 8 ft grounding rod.

For both the tower leg and guy wire grounding, be sure to use as short a run of wire as possible with smooth, even bends. Lightning hates sharp corners and may completely miss the ground rod if severe wire bends exist.

Figure 7-25. A well completed job! While a Whisper 175 captures the wind, a PV array complements the production of energy for this rural off-grid home. (Southwest Wind Power)

 Sun-spiration

The Great Blackout of 2003 started like many others for Ray Thompson and his family.....

"As night was falling and it became clear that the blackout was going to last more than a few minutes, I started to collect flashlights, candles and matches. As I was rummaging around for things I noticed the flicker of lights outside the living room window.

With a grin I popped outside and started plucking our solar powered yard lights from the pathway. The appearance of these lamps was similar to fireflies caught in a jar on a hot summer night. Each of the girls was given one and a few more were strategically set around the house. Candles appear to be obsolete now that renewable energy is here to help!"

Ray Thompson

Chapter 8
Micro-Hydro Electricity Production

If you are lucky enough to have a stream, river or waterfall on your property, it is well worth considering micro-hydro electricity production. While PV arrays are undeniably the simplest renewable energy source to install and maintain, micro-hydro systems (units under 1 kW capacity) can be the least expensive. The density of water is considerably higher than the wind. This increase in density allows a hydro turbine to be many times smaller than a wind turbine of similar output.

Smaller sized turbines and fairly simple installation help keep the costs to a reasonable level. In addition, hydro sites are generally able to supply electricity 24 hours per day. A smaller amount of power supplied over a longer period equates to a large amount of energy. For example, a typical PV array may be rated at 1,000 watts (W) of output power. Over a five-hour day of sunlight, this would equate to 5 kilowatt-hours (kWh) of energy. A small 200 W hydro turbine operating 24 hours per day would produce approximately the same amount of energy. On the downside, suitable hydropower sites are not as common as either PV or wind sites.

Understanding the Technology

As we discussed in Chapter 2 – Energy 101, electricity is produced from the kinetic energy in moving water under the force of gravity. In a manner similar to electricity, the pressure or *head* and the flow determine the potential energy in water. The vertical distance that water must fall under the influence of gravity determines the head pressure. Head is measured either in feet or may be expressed in units of pressure, such as pounds per square inch (psi) or kilo-pascals (kPa). It is interesting to note that water standing in a vertical tube will exert a known force at the bottom of the tube at a ratio of 1 psi (6.9 kPa) per 2.31 vertical feet (27.7 inches / 0.7 m).

This allows head to be easily converted back and forth between vertical height and pressure, as necessary.

Flow relates to the quantity of water flowing past a given point in a given period of time, which is expressed as *volume*. Typical units are gallons of water per minute (gpm) or liters per minute (lpm).

When the flow and head are integrated mathematically, power is the result that can be expressed in horsepower or watts. Once the units of waterpower are converted to watts by the turbine's generator, we are back to using common electrical units of measure. Because head and flow are interrelated, a high-head site with a low

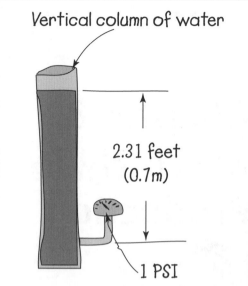

Figure 8-1. A vertical column of water 2.31 feet high (0.7 m) exerts a pressure of 1 psi (6.9 kPa).

Figure 8-2. This turgo wheel turbine and generator are suited for high-head and low-flow applications. This unit is installed in the Rocky Mountains of British Columbia and provides an output of 500 W. (Energy Systems and Design)

flow produces the same amount of power as a site with a low head and a high flow. For example, a site with 100 ft of head and 2 gpm of flow has the same potential energy as a site with 2 ft. of head and 100 gpm of flow. Most sites are located in an either/or situation. A high-head site is generally located in a very hilly or mountainous region, where the water might flow

Figure 8-3. A low-head turbine, such as this propeller-based model, requires high water flow. With 3 ft (1 m) of head, this unit produces 200 W of power. (Energy Systems and Design)

through several hundred to a couple of thousand feet (100 m to 1000 m) of plumbing line. Fortunately, a high-head site requires little flow, so that the pipe and turbine in this instance are fairly small.

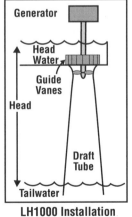

Figure 8-4 a, b. A low-head turbine is physically larger than its high-head cousin, but retains many of the same components. (Energy Systems and Design)

A low-head site offers the opposite issues. These sites are commonly known as "run of river" designs and can be sized to operate from 2 ft to 10 ft of head (0.6 m to 3 m). What we make up for in lower head, we repay by requiring more water flow and a larger turbine; nobody rides for free.

Water turbine types fall under two broad categories and dozens of sub-categories that we do not need to worry about. The main consideration with respect to turbine selection is whether there is a large flow of water and little vertical drop, or vice versa.

All turbines have a number of components in common, as do the civil works of the hydro site. Figure 8-4a and 8-4b show the workings of a low-head, high-flow hydro turbine manufactured by Energy Systems and Design of Canada (www.microhydropower.com). As the low-head turbine must have a large flow, the intake is formed using the large gray plastic "flume". Water enters the turbine heart and passes through a set of guide vanes. These vanes are stationary and are used to force the water to hit the *runner blades* at the correct angle. In this low-head turbine, the runner blades are shaped like a large boat propeller. As the water falls through the vertical head, the runner blades capture the kinetic energy and cause the generator to rotate, producing electricity. The draft tube forms the exit path for the water. The tapered draft tube acts a siphon, helping to draw water down the tube and away from the runner blades, increasing turbine efficiency. To increase the head and resulting output power of the turbine, a series of draft tube extensions can be added.

High-head turbines, such as the turgo wheel unit shown in Figures 8-5a and 8-5b, are considerably smaller, owing to the low flow of water required to turn the runner blade. Increased head and resulting water pressure make up for what is lost in flow. For example a turbine installed with 300 ft of head (91 m) will have a water pressure at the runner nozzles of 130 psi (900 kPa).

300 ft head ÷ 2.31 ft per psi = 130 psi nozzle pressure

Figure 8-5a shows a water intake pipe plumbed to a series of two shutoff valves, prior to entering the turbine. These valves may be used for turning off the turbine during routine maintenance or to lower water flow during dry periods. After leaving the shutoff valves, water is directed through a nozzle prior to striking the turgo style runner blade. Nozzle openings can be changed to adjust the water flow and resulting output power of the turbine.

The high-pressure water jet strikes the runner blade, imparting a turning effort or torque on the generator shaft, producing electricity. The water then exits the turbine heart by simply dropping away from the open bottom casting to return to the *tailrace* downstream.

The civil works comprise the dam, intake, penstock and tailrace sections of the hydro site. Figure 8-6 details the major components required for a high-head site. A dam structure provides a means of inserting an intake pipe and strainer midway

Figure 8-5 a, b. High-head turbines tend to be quite small, owing to the high water pressure and resulting low flow required to operate the tiny runner blade. (Energy Systems and Design)

between the riverbed and river surface. If water levels are sufficiently deep, it may be possible to eliminate the dam and simply install the intake as a submerged strainer or crib.

The dam's outlet pipe is fed to a stop valve that is used to drain the penstock system during servicing, or in the event of a broken pipe. Penstock pipes are often buried to prevent mechanical damage and chewing by animals. The penstock is also provided with an air valve at the top and a pressure relief and drain valve at the

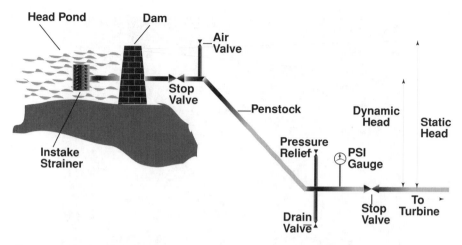

Figure 8-6. The dam, water intake, penstock and support components are known collectively as the civil works.

bottom. In addition, a second turbine stop valve and pressure gauge is also provided.

When the turbine stop valve is closed, water will stop flowing and the penstock will remain filled with water to the height noted as the "static" or "not moving" head level. As discussed earlier, the static head can be directly converted into a unit of pressure by dividing head in feet by 2.31 to produce psi. The pressure gauge will record this static pressure when the turbine stop valve is closed.

As the turbine stop valve is opened, water will begin to flow down the penstock to the turbine. As this occurs, the pressure on the PSI gauge will drop. The reason for this phenomenon is due to pressure loss from friction in the penstock pipe. The resulting pressure drop will lower turbine output accordingly. To circumvent this problem, a correspondingly larger penstock pipe may be used. Alternatively, lower friction polyethylene SDR pressure rated pipe may be chosen over PVC material. Pipe friction losses are noted in Appendix 9a and 9b. As pressure and head are equal, the flowing water pressure may be converted to "dynamic head". This refers to the "moving" or effective head level when the turbine is operating due to all losses in the hydraulic system.

If you have ever been in a home where suddenly turning off a water tap causes the whole house to shake, you have experienced the affects of *water hammer*. Water hammer is caused by moving water being suddenly stopped. Moving water contains kinetic energy (the energy of motion) that must be removed. If the turbine stop valve were suddenly turned off, the entire weight of water in the penstock must find some place to let off this energy. The effect is not much different than a freight train slamming into a solid wall; something has to give. Usually it's the penstock that gives. If the water hammer is severe enough, the penstock pipe will absorb the energy and possibly split or explode. To help stop damage to the penstock, a pressure relief valve is located near the bottom. Should pressure build due to water hammer, the pressure relief valve is sized to open before the pipe bursts.

An air valve is provided at the top most section of the penstock to eliminate any air trapped within.

Low-head or run-of-river sites are generally simpler, but still require a dam, intake and penstock pipe feeding an open flume, as shown in Figure 8-3.

What is a Good Site?

In order to determine if you have a suitable micro-hydro site, it is necessary to perform an evaluation of the available resources. The necessary variables for assessing a particular site are:

- water head level
- flow at the turbine intake
- penstock design issues
- electrical energy transmission distance

Water Head Level

Head may be measured using several techniques. One such example is as follows: A garden hose is held at the same level as the proposed intake from the stream. While one person ensures the hose remains filled with water, another person routes

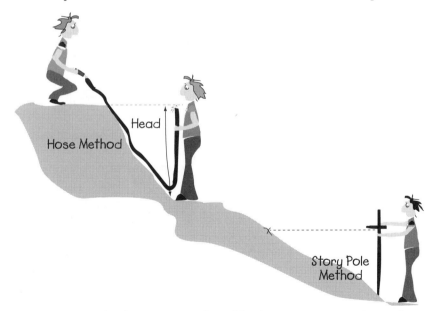

Figure 8-7. A garden hose or story pole and level are two common, low-tech ways of determining vertical distance on high-head sites. A GPS receiver or altimeter will also work.

the hose downhill, causing water to flow out. If the downstream end of the hose is slowly lifted, water will stop flowing out of the hose once the upstream and downstream ends are at an equal height. Measure the vertical distance from the point at which the water stopped flowing out of the downstream-end of the hose to the ground. This is the head measurement for the first section. This is repeated with the first person now holding the upstream end of the hose at ground level where the second person was standing. The process is repeated until the final head measurement is completed at the desired intake location for the turbine.

A variation on the hose method is to use a measuring stick or "story pole" and a carpenter's level, working uphill. The story pole is placed at the desired turbine intake location. A carpenters level is placed against the pole at any convenient height. By sighting along the level, a spot will be noted ('x' in Figure 8-7) which denotes a point of equal elevation. The story pole is moved to this spot and the process repeated until the stream intake location is reached.

A further and much simpler method is to use a global positioning system (GPS) receiver or accurate altimeter, if either is available.

Flow at the Turbine Intake

The easiest method to measure small flows is to channel the water into a pipe and using a temporary dam of rocks, plywood or whatever, fill a container to a known volume. By measuring the time it takes to fill the container you will be able to calculate the flow rate.

For higher flow rates the weir method is more versatile. This technique uses a rectangular opening cut in a board, which is set into the stream similar to a dam, as detailed in Figure 8-8. The water is channeled

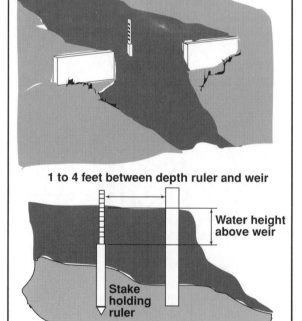

Figure 8-8. Measuring higher flow levels requires the use of a temporary dam or weir installed in the brook, stream or river.

into the weir and the depth is measured from the top of a stake that is level with the edge of the weir and positioned several feet upstream.

The resulting measurement can then be applied to the data contained in Table 8-1, which converts the weir measurement to U.S. gallons per minute of water flow. Multiply gallons per minute by 3.8 to determine liters per minute.

TABLE 8-1 - WEIR MEASUREMENT TABLE

Table shows water flow in gallons/minute (gpm) that will flow over a weir one inch wide and from 1/8 to 10-7/8 inches deep.

Inches		1/8	1/4	3/8	1/2	5/8	3/4	7/8
0		0.0	0.1	0.4	0.7	1.0	1.4	1.9	2.4
1		3.0	3.5	4.1	4.8	5.5	6.1	6.9	7.6
2		8.5	9.2	10.1	10.9	11.8	12.7	13.6	14.6
3		15.5	16.5	17.5	18.6	19.5	20.6	21.7	22.8
4		23.9	25.1	26.2	27.4	28.5	29.7	31.0	32.2
5		33.4	34.7	36.0	37.3	38.5	39.9	41.2	42.6
6		43.9	45.3	46.8	48.2	49.5	51.0	52.4	53.9
7		55.4	56.8	58.3	59.9	61.4	63.0	64.6	66.0
8		67.7	69.3	70.8	72.5	74.1	75.8	77.4	79.1
9		80.8	82.4	84.2	85.9	87.6	89.3	91.0	92.8
10		94.5	96.3	98.1	99.9	101.7	103.6	105.4	107.3

Example of how to use weir table:
Suppose depth of water above stake is 9 3/8 inches. Find 9 in the left-hand column and 3/8 in the top column. The value where they intersect is 85.9 gpm. That's only for a 1-inch weir, however. You multiply this value by the width of your weir in inches to obtain water flow.

Measuring the flow at different times of the year helps you to estimate your minimum and maximum useable water flows. If the water source is seasonably limited, you will have to depend on a hybrid system or fossil fuel generator to make up the loss of hydro energy.

You must also keep in mind that a reasonable amount of water should be left in the natural stream path to support life in that section of the stream. An intake screen will ensure that fish and other animals will not be driven into the turbine's intake system. When you have removed the energy from the water, it will be returned to the stream, with no loss in total water flow. Such an arrangement not only provides you with renewable energy, it is also environmentally responsible.

Zero Head Sites with Good Flow

After carefully explaining above that head and flow combine to make power, I am going to recant that statement enough to allow for an interesting class of zero head turbines. Zero head turbines rely completely on the flow or run of the river to extract their energy. There is currently only one commercially available unit of this class, sold by Jack Rabbit Energy Systems (www.jackrabbitmarine.com), which is shown in Figure 8-9.

Zero head units were originally designed to be pulled behind slower-moving sailboats as a means of charging batteries in these seriously off-grid systems. It stands to reason that if the turbine stops moving and the water doesn't, that the same power will be generated.

Figure 8-9. Zero head turbines, such as this model from Jack Rabbit Energy Systems, (www.jackrabbitmarine.com) feature special low-speed generators that can be mounted in fast-moving streams.

Installing a unit in a culvert or fast moving stream will start the out board motor like propeller spinning. Throw a ping-pong ball into the stream. If you can keep pace with the ball while walking, the river current is too slow. A brisk walk to partial jog will increase power output to 6 A at 12 V potential (1,700 watt-hours per day). If you need to increase your speed to a jog, the power output will also increase to its maximum of 9 A at 12 V (2,500 watt-hours per day).

These units are ideal where the stream will not freeze or for summer cottage installations.

Penstock Design Issues

All hydro systems require a pipeline. Even systems that operate directly from the dam require at least a short plumbing run or flume. It is important to use the correct type and size of pipe to minimize restrictions in the flow to the nozzle(s) or turbine runner blades. More power can be obtained from the same flow with a larger pipe, which has lower losses. Therefore, pipe size must be optimized based on economics. As head decreases, efficiency of the system decreases; it is important to keep the head losses low.

The pipe flow charts shown in Appendix 9 show us that 2 inch (50 mm) diameter polyethylene pipe has a head loss of 1.77 ft (0.5 m) per 100 ft (30 m) of pipe at a flow rate of 30 gpm (113 lpm). This is 17.7 ft (5 m) of loss for 1000 ft (300 m) of pipe. Using 2 inch (50 mm) PVC results in a loss of 1.17 ft (0.36 m) of head per 100 ft of pipe or 11.7 ft (36 m) for 1000 ft of pipe.

Polyethylene comes in continuous coils because it is flexible (and it is more freeze resistant). PVC comes in shorter lengths and has to be glued together or purchased with gaskets (for larger sizes).

Additional Pipeline Construction Details

At the inlet of the pipe, a filter should be installed. You can use a screened box with the pipe entering one side, or you can add a section of pipe drilled full of holes wrapped with screen or small holes without screening. Make sure that the filter openings are smaller than the smallest nozzle used on high-head machines. This prevents debris from blocking the nozzles and restricting power output.

The intake must be above the streambed so as not to suck in silt and should be deep enough so as not to suck in air. The intake structure should be placed to one side of the main flow of the stream so that the force of the flowing water and any debris bypasses it. This will also ensure that fish will not become entrapped in the intake system. Routinely clean the intake of any leaves or other debris.

If the whole pipeline doesn't run continuously downhill, at least the first section should, so that the water can begin flowing. A bypass valve may be necessary. This should be installed at a low point in the pipe.

For pipelines running over dams, the downstream side may be filled by hand. Once filled, the stop valve at the turbine can be opened to start the flow using a siphon action. If full pressure does not develop, a hand-powered vacuum pump can be used to remove air trapped at the high point.

At the turbine end of the pipeline, a bypass valve may be required to allow water to run through the pipe without affecting the turbine, purging the line of air or increasing flow to prevent freezing.

A stop valve should be installed upstream of the nozzle. A pressure gauge should be installed upstream of the stop valve so both the static head (no water flowing) and the dynamic head (water flowing) can be read.

The stop valve on a pipeline should always be closed slowly to prevent water hammer (the column of water in the pipe coming to an abrupt stop). This can easily destroy your pipeline, and for this reason, you may wish to install a pressure relief valve just upstream of the stop valve. This can also occur if debris clogs the nozzle.

Electrical Energy Transmission Distance

Let us assume we have a 12 V battery system and the transmission distance from the turbine to the battery bank is 200 ft (60 m) for the round trip. Because of electrical resistance losses in the transmission cable, we will require a slightly higher voltage at the generator to charge the batteries. If we assume that our turbine is providing 300 W of energy, this will translate into a current of approximately 20 A at 15 Vdc.

No electrical conductor is perfect, therefore a decision must be made involving how much energy should be lost in the wiring. Lower losses require larger, more expensive cable than one with slightly higher losses. An alternate consideration is to raise the battery voltage to 24 V or even 48 V. Remember that a higher voltage will result in lower current. The tradeoff is economy and is influenced by whether or not the battery system is already installed.

A good starting point is to work with a 10% loss. This yields a 30 W loss out of the original 300 W production assumption. The formula for conductor resistance loss when wire resistance is not known is:

Watts of loss ÷ Current ^{2} = Wire Resistance in Ohms

= 30 W loss ÷ 20 ^{2} Amps

= 30 W loss ÷ 400

= 0.075 Ohms of Wire Resistance

This is the calculated wire resistance that will produce a 10% electrical transmission loss. The wire resistance chart in Table 8-2 shows losses per 1000 ft of cable, while our example installation requires only 200 ft round trip.

1000 ft ÷ 200 ft x 0.075 ohms = 0.375 ohms per 1000 ft.

Table 8-2 indicates that #6 AWG wire has a resistance of 0.40 ohms per 1000 ft.

200 ft ÷ 1000 ft x 0.40 ohms = 0.08 ohms.

Increasing the wire size further reduces the losses in the same manner, albeit at a higher cost.

Wire Gauge	Diameter (Inches)	Ohms per 1000 Feet	Ohms per Kilometer
0000	0.460	0.05	0.16
000	0.410	0.06	0.20
00	0.364	0.08	0.26
0	0.324	0.10	0.32
2	0.258	0.16	0.52
4	0.204	0.25	0.83
6	0.162	0.40	1.32
8	0.128	0.64	2.10
10	0.102	1.02	3.34
12	0.081	1.62	5.31
14	0.064	2.58	8.43
16	0.051	4.10	13.39
18	0.040	6.52	21.33

Table 8-2. Copper wire resistance loss chart measured in Ohms per 1000 ft or 1 kilometer.

The power loss within a cable is shared between the current and the voltage. We must also take this into consideration. Voltage drop in the wire is equal to:

Current flow in amps x resistance of wire   = *Voltage drop in wire*
                                                = *20 A x 0.08 Ohms*
                                                = *1.6 V drop*

Therefore, if the battery voltage is 13.4 V, the hydro generator will be operating at 15.0 V. Keep in mind that it is always the batteries that determine the system voltage. That is, all voltages in the system rise and fall according to the battery's state of charge as will be discussed in *Chapter 9 – Battery Selection and Design*.

Site Selection Summary

After having determined the head, flow, penstock and electrical transmission issues, the question still remains; will this site be suitable for a micro-hydro installation? To finalize this decision, it is now necessary to refer to the turbine manufacturer's data sheets to determine how well the site will work with the available turbine selection. Hydro turbines possess a power curve data sheet in the same manner as wind turbines. Instead of relating wind speed to power, hydraulic units relate flow and head to generator output power

The power curve for the example model of turbine shown in Table 8-3 is based on the assumption that a minimum of 2 ft (0.6 m) must be provided before the unit develops usable power of 100 W output (or 2.4 kWh of energy per day). Even at this level of power output, the unit will require 500 gpm (1,900 lpm) of water flow. On the other hand, a high-head site will require much less water flow, but correspondingly higher head pressure levels.

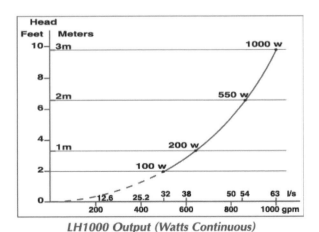

LH1000 Output (Watts Continuous)

Table 8-3. Output of a low-head turbine rated 1 kW at various flow and head levels. (Energy Systems and Design)

It will take a bit of juggling to determine which turbine will work best for your site, based on the turbine manufacturer's data. Additional turbine specific installation data is available from the various manufacturers listed in the resource guide in Appendix 3.

 *Sun-spiration*

Does using Renewable Energy make us better weather forecasters?

"As a gardener, I have always felt a wonderful connection with the earth and that I was in touch with the rhythms, patterns and cycles of nature. Being off-the-grid has strengthened this connection in surprising ways that have even improved my gardening.

Being so dependent on sunlight to make our electricity, I am more aware than ever of how many days of light that we have had, how many hours of sun to expect each season and particularly how much rain (and therefore cloud) we have had or not.

When my batteries are being filled to the point of overflowing, day after day, I am aware of how little rain has fallen and I am more aware than ever of how dry my gardens are getting!"

Michelle Mather

Chapter 9
Battery Selection
and Design

Why Use Batteries?

Why use a battery bank in the first place? This is a reasonable question with a simple answer. If you want to go off-grid now, there is no other way to store electricity as easily and economically. While it is possible to store electricity by making hydrogen or heat, the technology and cost benefit is completely uneconomic.

For folks who are considering a grid-interconnected system, there is an alternative. When your renewable system is generating more electricity than you are using, the system pushes excess energy into the electrical grid for consumption by your neighbors. When you require more energy than the renewable system can produce, you take electricity from the grid….a multi-trillion dollar battery of sorts. Do keep in mind that without any battery storage, when the grid goes out, so does the power in your house. The irony of this happening is almost sickening! For an added expense, it is possible to include a small battery bank into the equation, if only to have bragging rights amongst your neighbors.

So what about fuel cells? Automobile manufacturers are starting to use them, why not install some of these? Sorry, same problem. Fuel cells are wonders of technology and work well in specific applications, but home use isn't one of them. Besides, fuel cells don't store energy; they simply convert it from one form to another.

As their name implies, fuel cells require a source of fuel, usually in the form of hydrogen gas. When hydrogen and atmospheric oxygen combine inside the fuel cell, heat and electricity are the by-products. What we really want is an economical means of generating hydrogen from our surplus electricity. The stored hydrogen gas would then be fed back into a fuel cell to create the electricity we require on an as-needed basis.

Companies such as Electrolyser (www.electrolyser.com) make equipment to convert electricity and water into hydrogen, but the process is so inefficient that it is done only in very specific applications. Hydrogenics (www.hydrogenics.com) and Ballard Fuel Cell Ltd. (www.ballard.com) have been working on small fuel cell systems. Ballard, in conjunction with Coleman, produced a small home-sized unit to act as a battery backup system for computer and office equipment. Unfortunately, safety and supply issues with hydrogen gas killed that project for now. Come back in another 10 to 15 years and the story might be a little bit different.

Battery Selection

There are thousands of batteries available and the more you look the more selection and complexity you will find. Car, truck, boat, golf cart and telephone company batteries are a few examples that come to mind. Then, of course, there are the "special" kinds such as NiCad, nickel metal hydrid, lithium ion, and the list goes on and on. What is the correct selection for our renewable energy system?

First of all, let's wipe out a few of the legends and quick fixes that abound:

- Used batteries of almost any size and type are not the answer. It doesn't matter how good a deal you got, the reason the batteries are used is because there is a problem with them. All batteries die after a given life span or after being serviced too hard. Sure, they might work for a year, maybe even two or three, but lower charge capacity of older cells makes this a sour deal.

- Quality battery costs are high. A golf cart battery of the same capacity as a high-quality "deep-cycle" battery may cost less, but the reduction in life span is not worth the hassle except for very small or seasonal cottages. Changing batteries is backbreaking and dangerous work. You only want to do this once or twice in your lifetime, so stick with the best.

- Fancy batteries such as NiCad or lithium ion work well in cell phones and camcorders, for sure. But have you ever bought one of these babies? To get a lithium ion battery big enough to work in an off-grid system would probably cost as much as your entire house. Be thankful you can have one in your cell phone and leave it at that.

The only type of battery to consider in a full-time renewable energy system is the deep-cycle lead acid industrial battery. These batteries have been around a long time and have been engineered and re-engineered so that they offer the best value for money. Good quality industrial batteries should last between 15 and 20 years with a reasonable amount of care. In addition, the batteries are also recyclable, and many companies offer trade-in allowances for their worn out models, which also eliminates an environmental waste issue. The same is not true for NiCad or other exotic blends which are considered hazardous waste. Lead acid batteries are also available in numerous formulations, including spill-proof gelled electrolyte which is suitable for stomach-turning marine applications.

Figure 9-1. These models from Rolls Battery Engineering (www.surrette.com) are typical of long-life renewable energy system batteries. The two batteries in the foreground have the same voltage rating (6 Vdc). The model on the right has twice the capacity as the model on the left. (Surrette Battery Company)

How Batteries Work

The typical off-grid home battery uses a lead-acid composition, similar to a car battery. A single cell is comprised of a plastic case that holds a grouping of lead plates of slightly different composition. The plates are suspended in the case, which is filled with a weak solution of sulfuric acid, called *electrolyte*. The electrolyte may also be manufactured in a gelled form, which prevents spillage. The lead plates are then connected to positive and negative terminals in exactly the same manner as

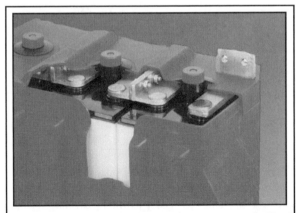

Figure 9-2. All lead acid batteries are comprised of lead plates suspended in a weak solution of sulfuric acid. The size of the plate and acid capacity directly affects the amount of electricity that can be stored. Each cell of the battery can be interconnected with others, increasing capacity and voltage. (Surrette Battery Company)

the AA and C cells described in Chapter 2. A single cell comprising one negative and one positive plate has a nominal rating of 2 V. The nominal voltage level changes as a function of how much energy is stored in the cell.

Connecting a load causes sulfur molecules from the electrolyte to bond with the lead plates, releasing electrons. The electrons then flow from the negative terminal through the conductors to the load and back to the positive terminal. This action continues until all of the sulfur molecules are bonded to the lead plate. When this occurs, it is said that the cell is discharged or dead.

As the cell is discharged of electrical energy, the acid continues to weaken. Using a device called a hydrometer (shown in Figure 9-3), we can directly measure the strength or specific gravity of the battery electrolyte. A fully charged battery may have a specific gravity of 1.265 (or 1.265 times the density of pure water). As the battery discharges, the specific gravity continues to drop until the flow of electrons becomes insufficient to operate our loads.

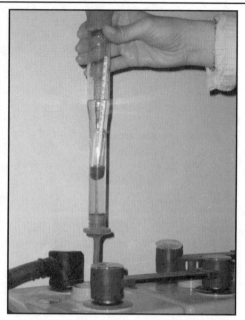

Figure 9-3. The hydrometer measures the density or specific gravity of fluids, such as the electrolyte in this cell. The higher the reading, the more electrons are stored in the cell, indicating a higher state of charge.

When a regular AA or C cell is discharged, the process is irreversible, meaning the cell cannot be recharged. Discharging an off-grid battery bank is reversible, allowing us to put electrons back into the battery, thus recharging it. Forcing electrons into the battery causes a reversal of the chemical discharge process. When a photovoltaic panel is placed in direct sunlight, it generates a voltage. When the voltage or pressure at the PV panel is higher than the battery, electrons are forced to flow from the panel into the cell's plate. Electrons combine with the sulfur compounds stored on the plate, in turn forcing them back into the electrolyte. This action raises the specific gravity of the sulfuric acid and recharges the battery for future use. Although there are many types of battery chemistry available, the charge/discharge cycle is similar in all types.

Depth of Discharge %	Specific Gravity @ 75° F (25° C)	Cell Voltage
0	1.265	2.100
10	1.250	2.090
20	1.235	2.075
30	1.220	2.060
40	1.205	2.045
50	1.190	2.030
60	1.175	2.015
70	1.160	2.000
80	1.145	1.985
90	1.140	1.825
100	1.130	1.750

Table 9-1. Using the hydrometer (shown in Figure 9-3), it is possible to accurately determine the state of charge, specific gravity and voltage of each cell in a battery bank.

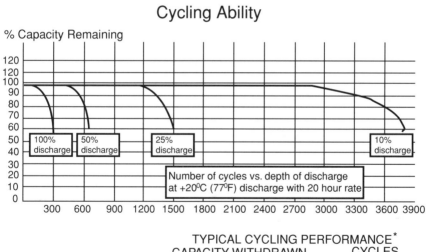

Figure 9-4. This graph relates the life expectancy in charge/discharge/charge cycles to the depth of discharge level.

Depth of Discharge

A deep-cycle battery has been given this name as it is able to withstand severe cycling or draining of the battery. A car battery subjected to more than a couple of "Oops, I left my lights on. Can you give me a boost?" mistakes will destroy the battery. Whereas a deep cycle battery may be subjected to much higher levels of cycling.

Figure 9-4 graphs the relationship between the life of a battery in charge/discharge/charge cycles and the amount of energy that is taken from the cell. For example a battery that is repeatedly discharged completely (100% depth of discharge) will only last 300 cycles. On the other hand, if the same battery is cycled to only 25% depth of discharge, the battery will last 1,500 cycles. It stands to reason that a bigger battery bank will provide longer life, albeit at a higher cost for the added capacity. A good level of depth of discharge to shoot for is a maximum of 50%, with typical levels of between 20% and 30%. Battery operation is discussed more fully in Chapter 14 – Living with Renewable Energy.

Operating Temperature

A battery is typically rated at a standard temperature of 75° F (25°C). As the temperature drops, the capacity of the battery drops due to the lower "activity" of the molecules making up the electrolyte. The graph in Figure 9-5 shows the relationship between temperature and battery capacity. For example, a battery rated at 1,000 amp-hours (Ah) at room temperature would have its capacity reduced to 70% at -4° F (-20° C), resulting in a maximum capacity of 700 Ah. If this battery is to be

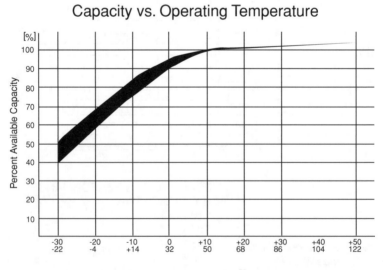

Figure 9-5. This graph shows the resulting reduction in battery capacity as the ambient temperature is lowered. A thermometer should be used to correct the specific gravity reading of very cold or overly hot electrolyte.

stored outside where winter temperatures may reach this level, the reduction in capacity must be taken into account. Installing batteries directly on a cold, uninsulated cement floor will also cause a reduction in battery capacity. Be sure to use a wooden skid or other frame to allow room temperature air to circulate around the battery, maintaining an even temperature.

Freezing is another concern at low operating temperatures. A fully charged battery has an electrolyte specific gravity of approximately 1.25. At this level of specific gravity, a battery will not freeze. As the battery becomes progressively discharged, the specific gravity also falls, until the electrolyte becomes water at a reading of 1.00.

Battery Sizing

A single cell does not have enough voltage or pressure to perform useful work. As discussed in *Chapter 2 – Energy 101*, single cells may be wired in series (to increase voltage) and/or in parallel to increase capacity.

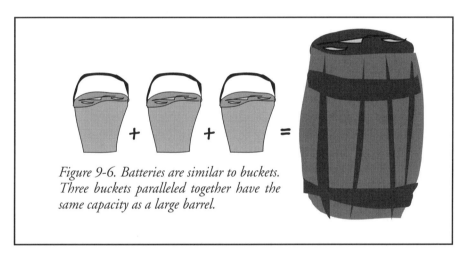

Figure 9-6. Batteries are similar to buckets. Three buckets paralleled together have the same capacity as a large barrel.

Batteries are just big storage buckets. Pour in some electrons and the "buckets" will store them for you. Take some out as you wish. If cells are wired in parallel, the capacity is increased. If the cells are wired in series, the voltage or pressure rises.

Batteries may be purchased in several voltages. Individual cells can be purchased which have a nominal rating of 2 V. The battery shown in the background of Figure 9-1 has two cells, each of which can be identified by the servicing cap on the top. Using interconnecting plates inside the battery case, the cells are wired in series providing 2 + 2 volts = 4 V total capacity. The batteries in the foreground have three cells, providing an output of 6 V. Using the same approach, a car battery contains six cells, providing a nominal 12 V rating.

Typical renewable energy system voltages are 12 V, 24 V and 48 V. Chapter 6 discusses the importance of selecting a battery with a voltage rating capable of meeting the amount of energy that is consumed in a day. It stands to reason that the more energy consumed, the larger the storage facility required holding all of this electricity.

Buckets are rated in gallons (liters) of capacity. We do not measure the amount of electrons stored in gallons, instead we use electrical energy units of watt-hours or amp-hours capacity. We discussed earlier that a battery's voltage tends to be a bit elastic. That is to say, the voltage rises and falls as a function of the battery's state of charge, as indicated Table 9-1. Additionally, when the battery is being charged from a source of higher voltage such as a PV ar-

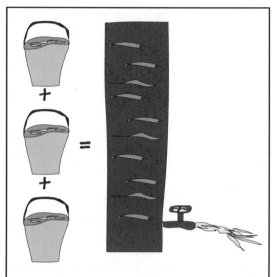

Figure 9-7. Batteries wired in series increase the voltage or pressure. In the same manner, a tall tank will shoot water further than a short tank due to the higher pressure.

ray in full sun, the battery voltage increases more. It is not uncommon for a 24 V battery bank to have a voltage reading of 29 V when undergoing equalization charging.

This fluctuation of battery voltage makes it very difficult to calculate energy ratings. You will recall that energy is the voltage (pressure) multiplied by the current (flow) of electrons in a circuit multiplied by the amount of time the current is

Application	System Size	Nominal Battery
A small cottage, hunting camp or ultra-conserver life style.	Up to 2 kWh per day	12 V
A full-time residence with careful energy consumption.	2 to 7 kWh per day	24 V
A large residence with adults and older children, having numerous appliances	Over 7 kWh per day	48 V

Table 9-2. The nominal battery voltage does not have any exact rules, but this table shows typical values for renewable energy systems.

flowing. The problem with this calculation concerning a battery is deciding which voltage to use.

Energy (watt-hours) = Voltage x Current x Time current flows

Battery manufacturers are a smart bunch. To eliminate any confusion with ratings, they have simply dropped the voltage from the energy calculation, leaving us with a rating calculated by multiplying current x time.

Battery Capacity (amp-hours) = Current x Time current flows

The assumption being that since the voltage tends to move, why try to shoot a moving target. If we really wish to look at our energy storage in more familiar watt-hour terms, it is a simple matter of multiplying the amp-hour rating by the nominal battery bank voltage as noted above.

Now, let's look at how this relates the to the storage level our battery must have in order to run our home. If you haven't already done so, it will be necessary to complete the *Energy Sizing Worksheet* in Appendix 7. This form will determine the amount of electrical energy you require to operate your home in watt-hours per day. Some typical consumption examples for small cottages to full-size homes are listed in Table 9-2.

For our sizing example, we will assume an energy consumption of 4,000 watt-hours (4 kWh) per day. This is a reasonable amount of energy for many off-grid systems. Grid-interconnected system users who wish to have a battery backup to ride through power failures will also have to calculate their required energy consumption using Appendix 7. In these cases, it may also be necessary to provide a separate wiring circuit for essential loads. For example, connecting house lights, a refrigerator and a few plugs to receive power during a grid interruption. Extra large or inefficient loads such as electric heating or air conditioning would remain off during the power failure. Providing this connection plan will add a small amount of cost, but will easily pay for itself in battery savings and longer running time during grid failures.

As the old saying goes: "Make hay while the sun shines." Unfortunately, renewable energy sources are subject to the vagaries of the weather. In the northeast area, November can be a bleak, dark, rainy month with little sunshine or wind energy production. A good rule of thumb is to have enough battery capacity to ride through a period of 3 to 5 days without *any* input from the renewable sources. With a 4 kWh consumption level per day, this would translate into a 20 kWh energy storage requirement.

4 kWh per day required x 5 days of capacity = 20 kWh capacity required.

Remember that we have to take into account the useable amount of energy stored in the battery bank, not the gross or rated amount. A maximum depth of discharge is typically 50%, with less being recommended. Most people would consider 5 days of no renewable energy input a worst-case scenario. Therefore a 50%

depth of discharge during this unusually poor production time is acceptable. This means that only 50% of the battery capacity is considered usable, requiring us to double the battery capacity rating.

20 kWh capacity required x battery de-rating factor of 2 = 40 kWh capacity

Some manufacturers will rate their batteries in watt-hours or kilowatt-hours capacity; however, most choose to use amp-hours as discussed earlier. To convert from watt-hours (or kWh) to amp-hours, divide the desired battery rating by the desired battery nominal voltage for your system, selected from Table 9-2.

40 kWh capacity required ÷ 24 V battery = 1,666 Ah capacity required

A quick scan of the resource guide data in Appendix 3 indicates that there are no batteries available with this much capacity. Several manufacturers provide batteries at 500 Ah. So, what to do?

Remember that batteries can be connected in series to increase voltage and parallel to increase capacity. Three sets of these 500 Ah batteries wired in parallel would provide a 1,500 Ah capacity. This is a bit below the ideal rating, but it's close enough. We could also consider 4 sets of 400 Ah batteries that would offer 1,600 Ahr. A bit of juggling may be required, but either way, we are in the ballpark. As discussed earlier, under-sizing the battery can lead to over-discharging and shortening of the battery's life. If you have to do any estimating or choosing of sizes, it is always in your best interest to round up. Just like your bank account, it never hurts to have too much money, or in this case, battery capacity on hand.

Each of the two rows of eight batteries shown in Figure 9-8 are shown in the schematic in Figure 9-9. Each row comprises 8 batteries. Each battery is fabricated with three cells as

Figure 9-8. Each shelf contains a group of 8 batteries with each battery containing 3 cells at 2 V/cell. A single string on each shelf is therefore rated at 48 V. Each of the two shelves of batteries are wired in parallel, thus doubling the capacity of the 48 V batteries.

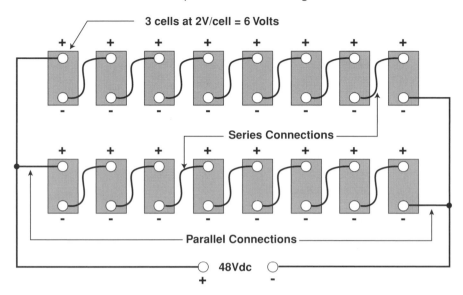

3 cells at 2V/cell = 6 Volts

Series Connections

Parallel Connections

48Vdc

+ -

Figure 9-9. The batteries in Figure 9-8 are shown here in a schematic. Each of the 2 rows x 8 batteries are wired in series, increasing voltage. The wired rows are connected in parallel increasing capacity.

noted by the three servicing caps on each battery. The nominal voltage of each battery is 6 V. When the batteries are wired in series by connecting successive "-" to "+" terminals, the voltages are additive. Thus, 8 batteries multiplied by 6 V per battery provide a total bank of 48 V.

Each of the batteries has a capacity of 500 Ah. Wiring the batteries in series does not increase the capacity. In order to increase capacity, the batteries must be connected in parallel so that they are both forcing electrons into the circuit at the same time. The parallel connection of the two strings of batteries is noted at the extreme right and left of Figure 9-9. As each battery string has a total capacity of 500 Ah, and two strings are wired in parallel, the total capacity is increased to 1,000 Ah.

Hydrogen Gas Production

Connecting a battery to a source of voltage will push electrons into the cells, re-charging them to a full state of charge. As the battery approaches the fully charged state, the electrolyte will start bubbling and hydrogen gas will be emitted.

Everyone knows the story of the Hindenburg dirigible: a few sparks and the hydrogen gas explodes into a ball of fire. Preventing this in a renewable energy system is actually fairly easy.

The battery bank should be accessible to allow for servicing and periodic in-

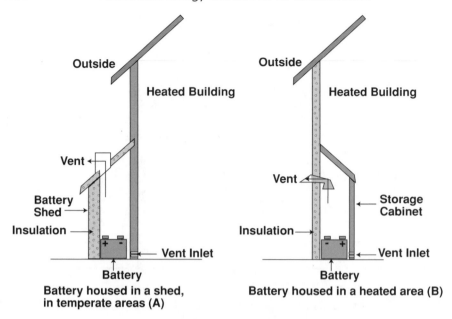

Figure 9-10 a and b. Batteries can be located outside of the living area where ambient temperatures do not deplete storage capacity (A). For colder climates, the batteries may be stored in a sealed box or closet (B). Both installations require a vent to the outside.

spection. However, safety is of primary importance. Storing them in a locked closet, shed or cabinet is a wise decision. The storage closet must be vented with a simple pipe that can be fabricated using a dryer vent or plumbing pipe (2 inch (50 mm) in diameter) and an insect screen. The screen should be located on the end of the pipe that is outdoors. Make sure the pipe slopes slightly upwards starting from the battery room to the outside. Hydrogen is lighter than air and will rise on its way outdoors.

For rooms with long or serpentine vent shafts, a power-venting fan can be installed as shown in Figure 9-11. A switch contained within the inverter is activated when the batteries reach a given voltage level operating the fans. This works because voltage is an indication of state of charge and hydrogen production, as discussed earlier.

An alternate method of controlling hydrogen out gassing is to install a set of hydrogen reformer caps (Hydrocap Corporation) such as those shown in Figure 9-12. These caps contain a catalyst that converts the hydrogen and oxygen by-products of battery charging back into water. The water simply drips back into the cell. The net effect is eliminated hydrogen gassing and lower battery water usage. Note that hydrogen reformer caps will not eliminate hydrogen outgassing. Venting must still be installed.

Safe Installation of Batteries

Large deep-cycle batteries are heavy and awkward. Some models weigh up to 300 lbs (136 kg) each. It is important that any frames, shelves and mounting hardware be of sufficient strength to hold them securely. Batteries should not be placed directly on cold cement floors as it reduces their capacity. During installation, make sure that you have adequate manpower to lift and place the batteries without undue straining. Tipping a battery and spilling liquid electrolyte (acid) all over you is very dangerous. Follow these rules to ensure a proper and safe installation.

- Batteries must be installed in a well-insulated and sealed room, closet or cabinet, preferably locked. There is an enormous amount of energy stored in the batteries. Children (or curious adults) should not be allowed near the batteries.

- Remove all jewelry, watches as well as metal or conductive anything. A spanner wrench or socket set accidentally set across two battery terminals will immediately weld in place and turn red-hot. This can also cause battery damage or a possible explosion.

- Hand tools should be wrapped in electrical tape or be sufficiently insulated so as not to come in contact with live battery terminals.

- All benches, shelves and the like

Figure 9-11. This battery room is power-vented. Installing small computer fans in the wall allows the room to be pressurized forcing hydrogen outside.

Figure 9-12. HydroCaps replace the standard vent caps on each battery cell. A special catalyzing material converts hydrogen and oxygen gasses given off during charging and converts them back into water.

must be strong enough to carry the massive weight of the battery bank. The batteries shown in Figure 9-8 weigh 4,800 lbs (2,200 kg), so use extreme caution when handling.

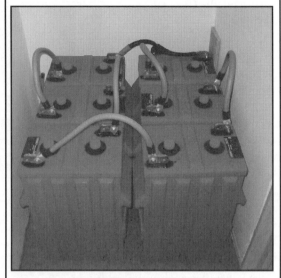

- Wear eye splash protection and rubber gloves when working with batteries. Splashed electrolyte can cause blindness. Wear old clothes or coveralls when working on batteries, and during routine maintenance. Electrolyte just loves to eat jeans.

Figure 9-13. This compact battery bank takes up a small amount of room in a utility closet. Rated 24 V at 1,300 Ah capacity.

- Keep a 5 lb (2.3 kg) can of baking soda on hand for any small electrolyte spills. Immediately dusting the spilled electrolyte with baking soda will cause aggressive fizzing, neutralizing the acid and turning it into water. Continue adding soda until the fizzing stops.

Battery and Energy Metering

Electricity is invisible, which makes it difficult to determine how much energy is actually stored in your battery bank. While it is possible to use nothing more than a hydrometer to read the stored energy level, this requires a fair bit of tedious fiddling and working with corrosive acid. A better approach is to install a multifunction meter such as the model shown in Figure 9-14.

These meters operate much like an automobile gas gauge, indicating how full or empty the tank is. Most meters provide dozens of features, with the most important functions being battery voltage measurement, energy capacity in amp-hours reserve, and the current flowing in or out of the battery. Another feature allows these meters to be mounted a long distance from the battery bank. You can read the battery capacity

Figure 9-14. A multifunction meter such as this Trace model by Xantrex (www.xantrex.com) takes most of the guesswork out of battery status monitoring.

from a convenient location in the house rather than running to the garage or basement.

Meters perform their work through a connection with a device known as a "shunt" (shown in Figure 9-15). The shunt is a high-capacity resistor that bypasses or shunts a small amount of the current flowing into or out of the battery to the meter. As the circuit is DC rated, the direction of current flow can be recorded. The amount of energy shunted is a calibrated ratio of the total amount of energy flowing in the battery circuit.

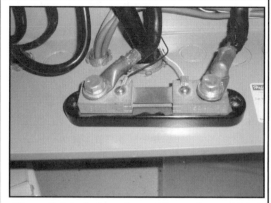

Figure 9-15. A shunt is precision resistor which is wired in series with the negative lead of the battery. A small current in proportion to the battery charge or load current is "shunted" to a multifunction meter such as the one shown in Figure 9-14.

Increasing the current flow into or out of the battery circuit causes a similar smaller calibrated current to flow to the meter.

By constantly monitoring the flow in and out of the battery, the meter keeps track of the battery energy level in amp-hours and percentage full. Well, it almost keeps track.... While there is no doubt that energy meters are very useful and accurate devices, they are not perfect. Batteries are not perfect either. The process of converting electricity into a chemical storage means in electrolyte is not 100% efficient. During the charge and discharge cycling of the battery, some of the electrical energy we put in is lost due to conversion inefficiencies. Worse yet, this inefficiency is not the same for adjacent cells, nor does the inefficiency level remain constant over the life of the cell. Due to this error, the energy meter will require recalibration once in a while when the battery is at a known fully charged state after measurement with a hydrometer. It is interesting to note that electrolyte specific gravity and battery voltage are directly related by this simple formula:

Battery Voltage = Electrolyte specific gravity + 0.84

For this equation to be accurate, the batteries should be under light or no load, with the electrolyte near room temperature. It is also important to perform this measurement approximately two hours after charging the batteries to ensure that gas bubbles in the electrolyte do not lower the specific gravity reading. The only way of absolutely knowing the battery energy level and general "health" is to use a hydrometer and measure the specific gravity. A thermometer should be used to correct the specific gravity reading of very cold or overly hot electrolyte according to the manufacturer's ratings. Correlate the corrected specific gravity to the data in Table 9-1 to determine the actual state of charge.

Batteries are charged a few times a year to a known level by performing an equalization charge, using the renewable sources or backup generator. When the batteries reach their known fully charged state, the meter may be reset, calibrating the battery level with the meter reading for the next couple of months. Equalization charging will be covered in Chapter 14 – Living with Renewable Energy.

In addition to the "main" features described above, metering may also contain some or all of the following "secondary" features, which aid in reminding when servicing functions should be carried out.

- number of days since batteries were fully charged
- total number of amp-hours removed/input to batteries since installation
- time to recharge battery indication
- batteries voltage is too low for proper operation
- battery is fully charged indicator
- battery is charging indicator

Figure 9-16. Battery installations should be neat and tidy and performed with the necessary tools to measure electrolyte specific gravity and temperature on hand. A clipboard and written record of each cell's "health" will ensure long battery life.

Summary

In this chapter we have discussed battery selection and safe installation. Chapter 10 – DC Voltage Regulation discusses the equipment necessary to ensure proper charging and regulation of the battery bank. Future chapters deal with the specifics of interconnecting cells and safe wiring practices as well as how to maintain your batteries while living with renewable energy.

"On August 14th, 2003, I was working on the layout of The Renewable Energy Handbook for Homeowners. At about 4:10 pm, approximately 50 million people in the northern United States and Canada lost their electricity, and became very aware of just how dependent they were on it.

I worked until about 6 pm and then went inside for dinner. We enjoyed a wonderful meal and conversation with friends, unaware that anything was wrong. By 8:30 pm the lights were on, our pump was supplying water to our faucets, our fridge chugged along keeping our perishables cool, and our daughters were watching television being beamed to us by a solar powered satellite.

Upon completion of their favorite show, they came into the kitchen to inform us that something seemed to be happening in the city. Leave it to teenage girls to know that parents with our passion for energy independence would hijack the television for the evening. Which we did in fact do, but after watching the chaos on CNN and CBC Newsworld for a few minutes, I went down to our elderly neighbors to run their generator for their fridge and freezer and to give them some lights so they could continue their bridge game.

Those in the know suggested that this was long overdue, and that the electricity grid has been stretched to the max for a long time. If this is the future of grid power, I won't gloat about not being dependent on it, but I will enjoy the comfort level that producing your own power brings. Plus I'll keep more food and wine on hand, in case the next black out runs longer, and our city friends are looking for a place to come and have a shower."

Cam Mather

Chapter 10
DC Voltage
Regulation

A charge controller is an important part of your renewable energy system as it provides regulation of the generated voltage. Charge controllers also provide protective functions, ensuring the batteries are not undercharged or overcharged, thereby prolonging their life.

Grid-interconnected systems not equipped with batteries do not require any form of charge controller as all of the available electrical energy is pumped directly into the grid at a fixed voltage. If your system will be grid-connected without batteries, feel free to skip this chapter.

All renewable energy systems generate more energy than can be reasonably used in the home. Even small systems will generate excess energy if no one uses the cottage. As a battery's state of charge increases, the voltage also increases. The renewable energy source produces power at a high voltage in order to provide sufficient "pressure" to cause electrons to flow into the battery. We learned in *Chapter 6 – Photovoltaic Electricity Generation* that a typical PV panel will produce 17 volts (V) for a 12 V nominal battery system.

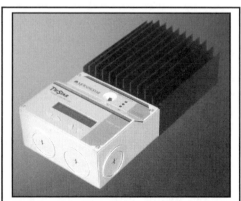

Figure 10-1. The charge controller ensures that your battery bank is properly charged and is operating within given operating parameters. (Morningstar Corporation)

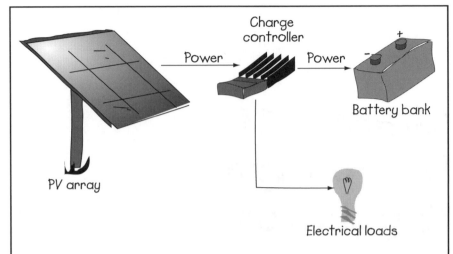

Figure 10-2. The series connected charge controller is able to connect and disconnect the PV array from the battery. This action limits charging current and voltage to safe levels, prolonging battery life.

The 5 V difference in pressure allows electrons to flow from the higher voltage source to the lower one, charging the battery bank.

As the battery bank fills its voltage rises, because it is not able to absorb any more energy. When nearing the fully charged state, the battery starts to outgas hydrogen and heat. This condition is known as overcharging, and if it continues for prolonged periods, damage to the battery may occur. This is where the charge controller comes in.

Charge controllers are available in two distinct categories known as *series controllers* and *diversion controllers*. Small cottage systems and PV-only charge controllers typically use the series controller method. Larger PV systems and systems using wind or hydro almost exclusively use the diversion controller method.

Series Controller

As the name implies, the series controller is wired in series between the PV array and the battery bank, as shown in Figure 10-2. In this arrangement, the voltage output from the PV array is fed through the series connected charge controller before being supplied to the battery bank. The charge controller monitors the battery voltage. Provided the battery is below a fully charged condition, PV power is allowed to flow into the battery. When the battery voltage level is sufficiently high to indicate that the battery is full, the charge controller disconnects the PV array from the battery, stopping the charging cycle. As the battery is subjected to household electrical loading, the charge controller senses the drop in battery voltage and reconnects the PV array. Think of the series controller as an automatic light switch that turns the flow of current to the batteries on and off as required.

The series controller may also include an option to protect from over-discharging, which is a condition that reduces battery life. This is known as a *Low Voltage Disconnect* or LVD function. As illustrated in Figure 10-2, electrical energy from the battery is supplied to the loads via the charge controller. Using this connection arrangement, the charge controller can automatically disconnect the electrical load whenever the battery charge state drops below a preset voltage point. An example of a charge controller with integral LVD circuit is shown in Figure 10-3.

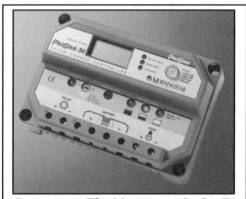

Figure 10-3. The Morningstar ProStar™ charge controller provides all of the charging and protective functions for small- to medium-sized renewable energy systems. An integrated energy meter is also provided with this model. (Morningstar Corporation)

During the night, the charge controller performs no functions other than electrically isolating the battery bank from the sleeping PV array. The reason this is done is to prevent energy flowing from the batteries back into the PV array. This condition *may* occur when the battery voltage is at a higher level than the PV array, which occurs at nighttime. The large surface area of the PV array absorbs a small amount of energy from the battery and dissipates it as heat. Over the course of a long winter night, this energy loss is measurable and can contribute enough inefficiency to become a concern.

Most PV module manufacturers incorporate "electrical check valves" or bypass diodes that prevent back flow from happening. Older, used or very small modules may not incorporate these diodes due to the small amount of loss attributed to them. For this reason, most series charge controllers contain a "PV array nighttime reverse-current protection" feature to eliminate the possibility of electrical back flow.

The main advantage of the series controller is simplicity, while the main disadvantage is wasting electricity produced by the renewable system when the battery bank is full. Small renewable energy systems often do not produce much excess energy, so simplicity is favored over capturing this small amount of excess energy.

Diversion Controller

The diversion controller method is shown in Figure 10-4. In this design, the renewable energy source is connected directly to the battery bank. All of the energy produced by the PV array or turbine can flow into the battery, providing a load connection at all times. This is an important distinction between the two designs of charge controller. Where a PV array can be connected and disconnected at will, many wind and hydro turbines must have an electrical load connected at all times. Removal of the load during operation would cause the turbine to accelerate to a

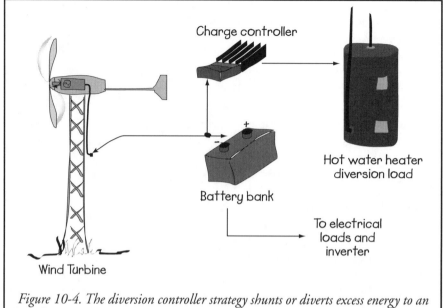

Figure 10-4. The diversion controller strategy shunts or diverts excess energy to an auxiliary air or water heater unit, based on battery voltage or state of charge.

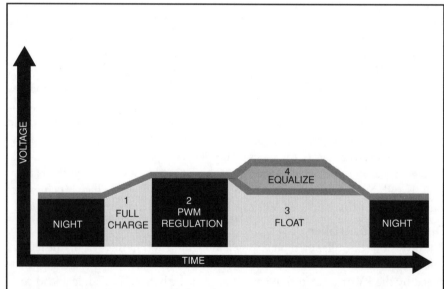

Figure 10-5. A charge controller regulates the amount of energy flowing into the batteries, ensuring a proper charge cycle. Quality charge controllers offer several charging modes. (Morningstar Corporation)

speed where damage may occur. A metaphor for this could be the way a car engine works: someone driving while holding "the pedal to the metal" results only in speeding tickets but no engine damage; however, doing the same with the car in neutral will likely cause engine damage as there is no load to limit engine speed.

To ensure a constant load for the turbine without overcharging the battery, the diversion controller will "shunt" or divert small bursts of energy to the diversion load. The diversion load in this example is a water heater, although an air heating or other dump load can be used. With this system, the diversion controller is able to divert all or part of the turbine's energy based on the battery bank voltage and state of charge, while maintaining a proper load level for the renewable source.

Charging Strategy

When the voltage is below a level that indicates the battery is not fully charged, all of the power from the PV array is allowed to pour in. As the battery voltage rises, the series charge controller rapidly cycles on and off (several hundred times per second), pulsing the PV power into the battery. In a similar manner the diversion controller rapidly cycles power to the diversion load. The net effect is to direct energy in a controlled manner, maintaining the battery voltage within preset limits.

To visualize this concept, go into a darkened room and flip on an incandescent lamp. The room will be lit at maximum brightness. If you rapidly flip the switch on an off, the bulb will alternate between a full and zero brightness state. The effect to your eyes (notwithstanding the flicker) is that the room is seen in a dimmer light. If the light switch could be controlled so that the switch spent more time on than off, the brightness would increase. Likewise, if the switch spent more time off than on, the brightness would decrease. Engineers term this rapid switching on and off *Pulse Width Modulation* or PWM.

To understand how the PWM concept is incorporated into a charge control system, let's take a look at a typical 24-hour day in the life of a battery bank. Figure 10-5 shows the effect of battery voltage over one complete night/day/night cycle. The graph shows time on the horizontal or "x" axis, while battery voltage is shown on the vertical or "y" axis.

• Nighttime

Starting at the left hand side of the graph, the battery voltage is shown at its lowest state at night when using a PV array for charging. During the night, the water pump may cycle, lights are used, the fridge operates and other loads draw energy from the battery bank. As we learned in *Chapter 9 – Battery Selection and Design*, the battery voltage will decrease as energy is removed from it. During this period, the charge controller prevents electrical back flow (discussed earlier) and waits patiently for the sun to rise.

• Sun Rise – Full or Bulk Charge

As the sun rises and shines on the PV array, electrical energy will start flowing and begin charging the battery. The battery voltage will start to rise based on the battery state of charge, as discussed in Chapter 9.

Until the battery state of charge reaches approximately 80%, all of the PV array power is applied directly to the battery, as noted in (1) of Figure 10-5. This charging condition is called "full or bulk charge" mode, and terminates when the battery voltage rises to 14.6 V for a nominal 12 V battery bank (consult manufacturers' data sheets for exact charger settings). For 24 V or 48 V battery banks, multiply the described settings by 2 or 4, respectively.

• Absorption or Tapering Charge

When the battery reaches approximately an 80% full state of charge, some of the energy is wasted in the form of "boiling" and out-gassing. Boiling is a slang term used to describe the breakdown of water in the battery electrolyte into its component elements, hydrogen and oxygen. This occurs when there is more energy applied to the battery than it can absorb.

Hydrogen is a gas that is lighter than air and very explosive when mixed with oxygen and an accidental spark, open flame or cigarette. For these reasons, batteries should be stored in locked and ventilated cabinets. It is possible to reduce hydrogen production while providing the optimum charging current to the battery by introducing a PWM tapering current, as shown in stage (2) of Figure 10-5.

During this stage, charging current is automatically lowered in an attempt to maintain the battery voltage at the bulk setpoint level of 14.6 V. Absorption charging continues for a defined period of time, which is typically one to two hours.

The excess energy produced during the absorption charging stage, which can be a considerable amount, is either wasted or diverted depending on the type of charge controller installed. A series controller will waste the excess energy. As this energy is non-polluting and often small in quantity, there is no concern. A diversion controller will shunt or redirect the excess energy to a dump load such as an electrical water or air heater, increasing the system's efficiency.

• Float Stage

Once the battery is fully charged, the charge controller will reduce the applied voltage to 13.4 V and begin the float stage, as shown in (3) of Figure 10-5. When the battery reaches this stage, very little energy is flowing into the battery, and nearly 100% of the renewable source's energy will be wasted or be shunted to the diversion load. Float mode will remain in effect until the battery voltage dips below a preset amount (typically 80% full state of charge) or until the renewable energy source stops producing power, which is typically at night. The process is then repeated for the next cycle.

• Equalization Mode

To visualize the equalization process, a simplified example is in order. Batteries can be thought of as buckets or drums that hold electrons. For this example, we will replace the electrons with water in our buckets. We can now assume that each bucket is the equivalent to one battery cell.

Over the course of a few dozen-charge cycles, energy is taken from and re-placed into the battery. This effect is the same as if I asked you to take 4 cups (1 liter) of water out of each "cell" or bucket. This represents a day's electrical load consumption. Assume that our PV array produced enough energy to put back 3 cups (750 ml) of "energy" back into the cells, leaving us with a deficiency of 1 cup (250 ml). This process is repeated with different amounts of water being added or removed over the course of a month or two, until such time as the batteries are back to a full state of charge. In our example, the buckets should also be full.

In reality, the water in each of the buckets is not at exactly the same level. A certain amount of spillage and uneven amounts of removal of water will leave buck-ets with varying amounts of water in them. The exact same scenario plays out in your battery bank. Over time there is a gradual change in the state of charge be-tween the cells, which can be easily noticed by comparing one cell's specific gravity with the rest of the cells' specific gravity.

If the difference is allowed to continue for an extended period, it can lead to the cell with the lower state of charge failing prematurely, as well as reducing the amount of available energy stored in the battery bank.

To correct this situation, a periodic (typically once per month) controlled over-charge is conducted, which is known as an equalization charge. This is illustrated by step (4) in Figure 10-5. Chapter 14 – Living with Renewable Energy discusses how to determine when equalization is required.

Equalization charging is normally conducted early in the morning on a sunny or windy day. The charge con-troller is set to equalization mode and the normal bulk (step 1) cycle is com-pleted. In equalization mode, the con-troller raises the battery voltage to 15.5 V (step 4). Equalization mode is main-tained typically for 2 hours, after which battery voltage is reduced with the charge controller entering float mode (step 3).

If we go back to our bucket exam-ple, the effect of equalization mode would be similar to using a garden hose to add water to the buckets and over-filling them. When you stop adding water, the buckets are topped up right to the rim.

Figure 10-6. Using large amounts of diverted electrical energy is easy if heat enters the equation. An air heating element, such as the model shown in this photo (available from www.realgoods.com), will absorb up to 1 kW of excess energy and provide some home heating to boot.

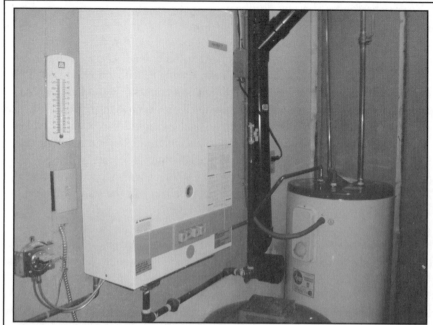

Figure 10-7. This home uses a high-efficiency, in-line gas water heater. The conventional storage water heater at rear is the diversion load, which absorbs waste energy by preheating the cold water fed to the gas water heater.

So where does the "extra" electricity go during the equalization mode? During this charging stage, the excess energy applied to the batteries will cause violent bubbling of the electrolyte, emitting large amounts of hydrogen gas and water vapor. It is necessary to monitor electrolyte levels prior to this charging stage to ensure the cell plates do not become uncovered.

If your batteries are equipped with hydrogen reformer caps, ensure they are removed during equalization mode.

Diversion Loads

A diversion load is described as any load that is large enough to accept the full power of a renewable source. During battery charging, and in particular the final float stage, a large percentage of the renewable energy system's output will be shunted to the diversion load.

Although the diversion load may simply waste the excess energy applied to it, a better approach is to put this juice to work. After all, you did pay for the PV array or wind turbine, so why waste the "excess" energy other people have to pay for? Using the excess energy can be a tall order. Suppose you have a 2,000 W PV array, a 1 kW wind turbine and both operate at maximum output for 5 hours while you

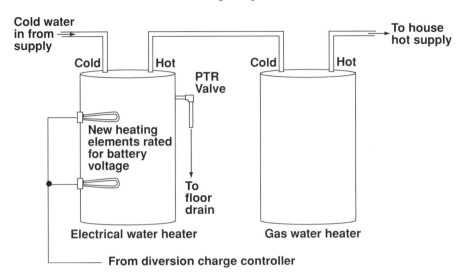

Figure 10-8. A diversion water heating load is connected in series with a standard gas water heater. The electric water heating elements are replaced with others rated the same as the battery bank nominal voltage and are connected to the diversion charge controller.

and your family are on vacation.

2 kW PV array + 1 kW wind turbine x 5 hours = 15,000 Watt-hour production

Assume for a moment that the house loads are zero (no fridge or lights on). What can you do with this energy? 15 kWh of production is a lot of energy. You could operate a thousand 15 W compact florescent lamps for one hour or ask your neighbors to have a really big party at your house.

A more common solution is to use the energy for heat, electrical heat. You will recall that large heating loads do not belong in an off-grid house. This is true, when the load is a *required* part of your day-to-day living. On the other hand, excess electricity is just that, excess. This free energy can be used in many efficient ways to help offset other systems that cost money to operate. Room and water heating are two very common solutions. If all of your waste energy is produced during the swimming season, consider dumping the energy into a hot tub or swimming pool. Just remember that the intended load must not be counted upon as part of your overall energy budget. If the weather cooperates and you make more energy than you can use, consider it a windfall.

The most common diversion loads are air and water heaters such as those shown in Figures 10-6 and 10-7. For PV-based systems, the majority of excess energy is produced during the summer months when air heating is not required. Wind systems tend to provide maximum power during the late winter and spring periods. Consider the choice carefully. No one wants an air heater cooking along when the mercury is in the 90s. My personal preference is using a water-heating load, as shown in Figure 10-7. This arrangement requires a bit more installation work, but the wasted energy will actually return dollars by offsetting propane used for domestic water heating.

In this system, a standard electric storage water heater is purchased and the 240 V heating elements removed. New elements are installed which have the same voltage rating as your battery bank. The elements are then wired to the diver-

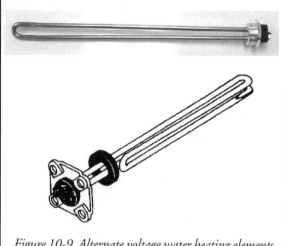

Figure 10-9. Alternate voltage water heating elements are available from numerous sources, including Real Goods at www.realgoods.com.

sion type charge controller. The water heater is plumbed so that cold water flows into the electric heater and out to the propane model, as depicted in Figure 10-8.

During normal operations, the cold water supply enters the electric water heater. If there is no excess energy, the cold water will absorb some room heat, capturing a small amount of supplementary energy before heading to the gas water heater. As this incoming cold water is below the setpoint temperature, the gas heater will supply the necessary energy to meet demand.

During periods of high-energy production, the diversion charge controller will feed excess energy to the electric water heater. This excess energy may heat the water to the setpoint temperature or beyond. Feeding preheated water into the gas heater reduces or eliminates the need for any further heating, saving you propane and dollars.

It is very common for the water in the electric heater to rise above the setpoint temperature, particularly during travel or vacation periods, which often tends to be in the energy rich summer months. Safety (and building codes) dictate that a Pressure and Temperature Relief Valve (PTR) be installed on the tank. The outlet pipe should be run to a floor drain or other suitable exhaust to allow very hot water to be safely drained away in the event the water tank temperature rises to an unsafe level.

For the same reason, your plumber should install a buffering valve. These valves ensure that the hot water supply delivered to the plumbing fixtures is within a comfortable and safe temperature range.

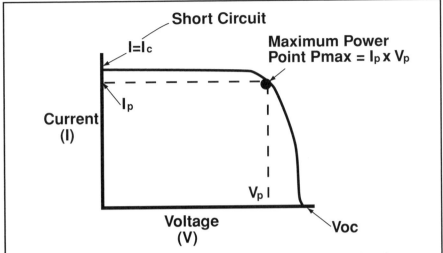

Figure 10-10. The Maximum Power Point (MPP) for a PV module is the point at which current and voltage outputs provide the maximum amount of power (Power = Voltage x Current).

Maximum Power Point Tracking

A word is required about *maximum power point trackers* or MPPT controllers. This technology, which is new to the renewable energy market, is now available in controllers manufactured by RV Power Products (www.rvpowerproducts.com). They have several models to choose from, all of which offer the charge control functions described above.

The MPPT function is a little bit difficult to understand, but well worth spending a few moments to review. All PV modules produce power based on the product of the voltage and current. This can be expressed by multiplying the rated current by the rated voltage at load, delivering the rated power of the module. This is shown in graphical form in Figure 10-10. The output voltage of a module is shown on the "x" axis with the output current on the "y" axis. In Chapter 6, we discussed how a PV module would generate a very high open circuit voltage when not connected to a load (i.e. no current is flowing). This point on the graph is noted at position Voc (Voltage Open Circuit), also indicating that there is zero current flowing. The resulting power output is obviously zero.

As a load is connected to the panel, increasing current causes the voltage to slowly drop. A short circuit (directly connecting the "+" to "-" terminals together) will result in high current levels, but the voltage will approach zero again, thereby producing no useful power.

The most efficient point on the curve occurs when the product of the voltage pushing and the current flowing produces the highest value. This occurs at one point, which is known at the Maximum Power Point MPP. Unfortunately, our battery voltage is unlikely to be located at this point. As we have learned, battery voltage is quite elastic so there is a high probability that the PV module output will only transit this point occasionally.

The MPPT controller monitors the battery voltage and flow of current, and determines where the MPP is for the given level of light illuminating the array. The effect of this technology is to increase the output current of the PV array by means of a simple direct connection, as used in series and diversion load controllers.

Testing shows that MPPT tracking systems increase the wattage of a PV array by an average of 15% to 30%. On a PV array pushing 800 W into a battery bank, this will provide an average increase of 120 W with a peak of double this figure. This is approximately equal to one or two modules of "free" power.

Figure 10-11. RV Power Products (www.rvpowerproducts.com) manufacture several models of MPPT trackers with built-in charge controllers.

There are downsides to this technology as well. An MPPT tracker is intended for use with PV arrays. As current models do not provide diversion load capabilities, you will need a second controller for wind or hydro turbines, which require a constant load. This loss of diversion energy must be calculated into the overall system efficiency. If the system increases battery charging efficiency then throws the excess power away when you could be saving water-heating energy, then the savings may be questionable.

For PV-only based systems, consider an MPPT controller as an alternative to an active tracking mount. You will have to compare cost and seasonal performance as well as load consumption patterns to determine which unit is suitable for your requirements:

MPPT is best when:
- winter loads are higher than summer
- the system is PV only
- there is no room for an active tracker

Active trackers are best when:
- summer loads are high, for example air conditioning or swimming pool operations
- the system is hybrid; PV/wind/hydro
- you have a wide "sun window" from early morning to late afternoon

Voltage Regulator Selection

Regardless of which voltage regulator design you select, it is important to match its electrical rating to your system.

The regulator rating is based on the maximum current that is allowed to flow into the batteries (series regulator) or into the diversion load (diversion regulator). For example, the model shown in Figure 10-11 is rated at 50 amps, maximum continuous rating. It will be necessary to add all of the peak charging/diversion currents expected in your system. Once this value is determined a derating or safety factor of 25% is added.

peak charging/diversion current x 25% safety factor = regulator rating in amps

Purchase the next largest regulator from this calculated rating.

For further details refer to *Chapter 13 - Putting it All Together Safely.*

 *Sun-spiration*

"I remember when the ice storm settled in and all around us was the eerie sounds of trees cracking and snapping in the surrounding woods.

Bill and I sent word we would help wherever we could. My Mom decided if we delivered coffee and some drinking water each morning she would prefer to stay and watch over her house. She had lots of company dropping in keeping the kerosene lamps lit and wood stoves burning away.

Bill's Mom decided it would be a whole lot warmer and brighter at our house, so she packed a bag and made her way over with a neighbor and her dog Cindy.

We also had Margaret come over from the senior's housing complex in the local town of Clayton. They were given short notice to leave the premises for security and safety reasons. Margaret's family was not close by and we were happy to have her over.

Once we had all the ladies settled in, we started cooking food that everyone had brought over and could not leave in the fridges and freezers. We invited neighbors to come over for a hot meal and watch the news. Everyone felt so lonely and out of touch without any power. It was fun to watch all these people glued to the TV set, crowded in our living room. You could have heard a pin drop, everyone was so quiet and focused on the latest situation surrounding the ice storm damage.

We had our nieces sleep over as well. Their parents were worried about them smelling kerosene from the heaters they were using to try and keep their house from freezing. We had regular scheduled bath times for many of the close neighbors. I could hear my sister Nancy sigh as she soaked in a nice hot bubble bath one night!

Bill and I would cuddle up on a futon on our living room floor with our two dogs lying beside us. I would think of those who didn't have power and hope their troubles would soon be over. I would also think of the people sleeping in our beds upstairs and smile to myself that we were helping to make someone's life just a little bit easier.

I felt good inside when I drifted off to sleep and realized just how self-sufficient our home is."

Lorraine Kemp

Chapter 11
DC to AC Conversion
Using Inverters

In earlier chapters, we have dealt with renewable energy sources producing electrical energy in a direct current (DC) form from a PV array. Wind and hydro turbines generate alternating current (AC), but convert it back to DC for storage in battery banks.

This chapter deals with how to convert DC power back to AC power for use with our electrical appliances in off-grid systems, as well as "selling" excess energy in grid-interconnected systems.

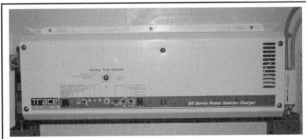

Figure 11-1. The renewable energy world would be lost without high quality inverters such as this Trace model from Xantrex Technology Inc. (www.xtantrex.com).

In the last decade, advances in power technology components and low-cost microcomputers have led to a revolution in inverter design. In the "old days", if you wanted to build a renewable energy system, it could only be accomplished using 12 Volt (V) direct current appliances similar to those used in recreational vehicles. Grid interconnection of renewable energy was as far away as the idea of the Internet. Times change, the Internet is here and so are very high quality inverters for both on- and off-grid applications.

The basic inverter is a device that takes low-voltage DC power and converts it to AC power then "steps up" the voltage to match domestically supplied power from your utility. In practice, many inverters offer a whole host of additional features and functions. Among the extras supported within the inverter are:

• battery charging capability
• transferability of house power between a generator and the inverter
• low and high-voltage alarms and disconnection (LVD function) of the battery bank
• energy savings – sleep mode turns inverter "off" but comes back on at a flick of the first house light switch
• automatic start and stop of a backup generator
• maximum Power Point Tracking for grid-interconnected systems
• full safety protection for both homeowners and utility workers

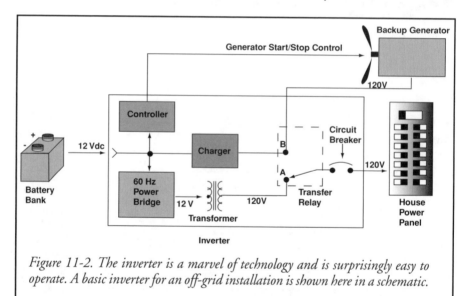

Figure 11-2. The inverter is a marvel of technology and is surprisingly easy to operate. A basic inverter for an off-grid installation is shown here in a schematic.

An inverter connected for off-grid operation is shown in a schematic in Figure 11-2. Energy stored in the battery is directed to the inverter's internal components. A power supply and controller determine the sequence of events required to make the unit function as desired. When the inverter is activated, the controller starts a high-powered oscillator or power bridge, which generates AC power at the voltage of the battery bank: typically 12 V, 24 V or 48 V.

The low-voltage AC power is set at a frequency of 60 cycles-per-second or 60 Hertz, for the North American market. Inverters in Europe and Asia operate at 50 Hertz. You will recall from *Chapter 2 – Energy 101* that 60 Hertz means that the polarity of the voltage is switched back and forth 60 times-per-second. Imagine a simple "C" cell battery is inserted into a flash light first one way then the other 60 times-per-second, as shown in Figure 11-3. If we plotted the polarity of the battery each time it was inserted into and out of the battery clips shown in "A", we would obtain a plot shown in "B".

The advantage of AC power over DC power is that the voltage can be easily stepped up or down using a transformer. A transformer inside the inverter receives the low-voltage AC power and steps it up to a utility standardized 120 V or 240 V. In our example inverter in Figure 11-2, the transformer is designed to accept a nominal 12 V AC input from the power bridge and convert it to 120 V output. The 120 V AC power output from the transformer then enters a transfer relay. The controller computer operates the transfer relay. When the inverter is operating in its "normal make AC mode", the transfer relay is connected between the inverter output, feeding 120 V or 240 V alternating current to the house's circuit breaker panel. From this

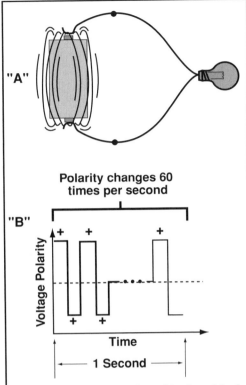

Figure 11-3. A battery flipped back and forth in its socket 60 times-per-second will supply alternating current to the light bulb load. The waveform at "B" shows the resulting plot of the changing polarity over a one-second interval.

point on, the wiring within the house is normal, with the inverter feeding the *mostly* unsuspecting appliances.

Inverters can detect when the battery bank is a bit low and can be programmed to shut down the inverter and sound an alarm. This prevents inadvertent over-discharging of the battery bank that will shorten the life of the batteries. Slightly more advanced inverters will cause the internal controller to send a start command to the backup generator. When the generator is confirmed to be running, the inverter switches the transfer relay so that power from the backup generator is fed into the house panel. This transfer occurs in the wink of an eye and barely causes a "flick" of the lights. Once relieved of inverter duties, the unit will draw power from the generator through the charger system, replenishing the batteries. It should be noted that all inverters with a built-in charger use the same multi-stage battery charging technique as is described in Chapter 10.

When the batteries are full, the inverter reverts to "float charging" mode until the generator is manually turned off. In advanced units, the inverter will stop the generator, switch the transfer relay back to "normal" mode and begin supplying the house power.

Inverter's AC Waveform

You may have noticed that I said earlier, "…feeding the *mostly* unsuspecting appliances." The electrical utility generates its AC power using mechanical generators connected to turbines. When a rotary generator creates an AC cycle, the waveform is sinusoidal, as shown in Figure 11-4 A. Compare this waveform to that of our simple inverter described above and shown in Figure 11-4 B. Each waveform has been magnified to show what is known as one cycle, being 1/60 of a second long. Obviously, 60 of these cycles occur in one second.

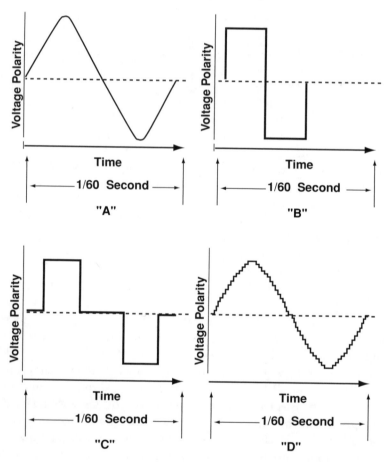

Figure 11-4.

The main difference between the two waveforms is that the utility generated one in "A" has the voltage ramping up slowly until it reaches a peak level, then the voltage falls back down slowly as the polarity reverses. By contrast, the simple inverter shown in "B" snaps quickly between points of opposite polarity, hence it is given the appropriate name "square wave". Only the very cheapest and oldest inverters operate using the simple square wave.

Improvements in technology led to the "modified square wave" (also known as a "modified sine wave", but that's really pushing it) as shown in waveform "C". The inverter shown in Figure 11-1 utilizes this format of waveform and it still retains a significant share of the inverter market today.

Inverters using a square or modified square wave will operate 98% of all modern electrical appliances without problems. However, the fast switching edges of the waveform can produce a buzzing or humming noise in some items such as cheap stereos, ceiling fans and record players. One primary advantage of this waveform is the ease of producing it. Inverters utilizing this pattern are very robust, electrically efficient and relatively inexpensive.

A fairly recent development in inverter technology is the sine wave model that outputs a waveform similar to that shown in "D". This digitally synthesized waveform shape is created using a technology similar to that used to record digital sound on a compact disk or MP3 file, and we all know how good those sound. In fact, today's inverter technology offers waveform distortion of less than 5% and frequency and voltage tracking that far exceeds even the most expensive utility power station quality.

Grid Interconnection Operation

A sine wave inverter that generates AC power of utility quality should be able to supply power to the grid. Ten years ago this was considered heresy by the electrical utilities. Times change and the demand of cleaner renewable energy (including political head knocking) has loosened the utilities' grip on its monopoly. Of course, some areas of North America are slower than others at getting the point, but the change is occur-

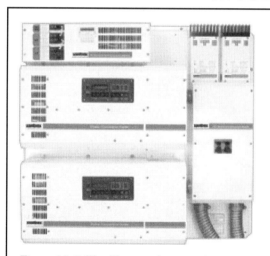

Figure 11-5. Two Xantrex sine wave inverters are "stacked" to provide both 120 V and 240 V AC power at better than utility quality power. Also shown are two model C60 charge controllers (upper right) and a DC circuit breaker (bottom right). (Xantrex Technology Inc.)

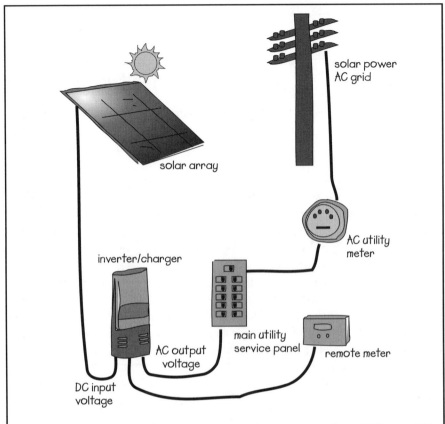

solar power
AC grid

solar array

AC utility
meter

inverter/charger

AC output
voltage

main utility
service panel

remote meter

DC input
voltage

Figure 11-6. A typical grid intertie inverter is shown connected to a PV array. This simple connection allows the electrical meter to spin backwards when energy is being exported (sold) to the grid. In cloudy times, you import (purchase) power from the grid as usual.

ring and inevitable. Take California for example: after the problems with rolling blackouts, sky-high energy costs and never-ending smog, the politicians got the message. Now anyone can install a bunch of PV panels on their roof, connect them all up through a grid intertie inverter and watch the electrical meter spin backwards.

The grid-connected system shown in Figure 11-6 is typical of a simple effective means to allow renewable sources to supply energy to the utility. In this arrangement, electrical energy flows from the grid, through the utility meter and into the home's power panel and electrical loads. This is the normal import or purchasing mode that most homes are familiar with. During importation of power, the meter records the power usage in a numeric format; as power importation continues, the meter's numbers increase.

When the sun starts to shine on the PV array or the wind or water turbine starts spinning, the inverter unit synchronizes its own sine wave to that of the utility. In order to export power, the inverter creates a waveform that has a higher voltage than the utility source, causing current to be exported to the grid. During power exportation the utility meter numbers decrease, reducing the recorded power importation numbers, thereby reducing the amount of money you owe to the utility. In effect the grid is acting as battery on your behalf.

For safety reasons, the inverter periodically checks the utility voltage to make sure that it is still functioning. If it is not, the inverter shuts down and waits until the grid has been running continuously for 5 minutes before re-applying power. This test eliminates the possibility of a condition known as *islanding*. Islanding can occur when power from the grid is knocked out by downed power lines resulting from a storm. If the inverter were to continue trying to export power to the grid and a utility worker unsuspectingly touched the downed wires, serious injury may occur.

Inverter Ratings

An important consideration when purchasing an inverter are the electrical ratings of the unit and whether or not you should purchase a "modified square wave" or "sine wave" model. Following is a list of items to look for when shopping for any inverter.

Sine Wave versus Modified Square Wave Inverter Models

As discussed earlier, the major issues between these models are cost and operating difficulty for very few electrical loads. Costs are always changing, with sine wave models starting to take the majority of the market share. Companies such as Xantrex, Sunny Boy, SMA and Outback all manufacture a wide variety of models to suit any budget. For a seasonal cottage with infrequent use, by all means use a modified square wave model. Costs are lower and the units are built like trucks and should last a very long time.

Running a full-time home with more electronics and the like, sine wave units are the way to go as long as you can budget one in. If not, don't worry; there are thousands of homes both past and present that use modified square wave units without any concern.

Output Voltage

North American homes are wired to the grid so that loads may be connected at 120 V or 240 V. Normal wall plugs are rated at 120 V, while heavy electrical loads such as electric stoves and furnaces, clothes dryers and central air conditioning units operate at 240 V.

If you wish to retain 100% standard wiring of the house panel, or if you are grid-connected, purchase a 240 V rated inverter or "stack" two 120 V models, as prescribed by the manufacturer.

Inverter Continuous Capacity

Inverter capacity refers to the amount of power that the unit can supply at one time continuously. You will recall that power represents the voltage multiplied by the current, which is expressed in watts of power. When you are at home, more than one electrical load is likely to be turned on. In order to determine the inverter continuous capacity you require, it is necessary to add up all of the electrical loads that are likely to be turned on at any given time. For example, assume that a well pump (1,800 W), washing machine (500 W), gas clothes dryer (400 W) and a bunch of CF lamps (100 W) are all on at the same time. The total power requirement of these loads is:

1,800 + 500 + 400 +100 W = Total Continuous Power
= 2,800 W

This means you will require an inverter that is capable of supplying at least 2,800 W continuously.

Remember to calculate this value when filling out the *Electrical Energy Consumption Worksheet* in Appendix 7. When you are shopping for an inverter, you will find that they tend to be sized in "building block" sizes. There are numerous small models below 1,000 W, though 2,500 W and 4,000 W are common sizes. It never hurts to have a little bit of extra capacity in case the kids need a bunch of power for their new garage rock band. (On the other hand, maybe limiting available power might not be a bad idea after all!)

When trying to read the continuous rating of an inverter, you may be confronted with the term "VA" rather than watts. VA refers to the voltage multiplied by the current that we have understood to mean wattage. At the risk of trying to split hairs, when an AC motor load is operated, there is an effect known as *power factor* that has to be taken into consideration. In the average home, VA closely approximates watts and may be used interchangeably. It is prudent to reduce the rating of the inverter by as much as 20% if you will be operating air conditioning units, pool and spa pumps or other similar high-wattage, *induction* motor loads. If you are operating a shop full of large wood working tools with many of the motor loads running at the same time, it would be best to allow for an inverter derating of approximately 25% to allow for power factor. In practice there is little that you can do other than to be aware and reduce simultaneous usage of motor-driven appliances or to purchase the next larger sized inverter. An additional point to note is that universal motors that have brushes (sparks may be seen when the motor is running) can be run safely without any derating issues. Examples of these motors include:

- regular and central vacuum cleaners
- food processors and mixers
- drills, routers, shopvacs, radial arm and circular saws
- electric chain saws and hedge trimmers

• electric lawn mowers

If in doubt, contact a motor rewinding or repair shop for clarification.

Inverter Surge Capacity

The surge capacity is an indication of how much short term overload the inverter will be able to handle before it trips on an overload condition. The reason this indication is necessary is to allow for some large, particularly motorized loads that require two to three times their running power to get started. Although this start period is very brief and lasts a fraction of a second, it should be considered. The main concern is whether or not you have any "unusual" electrical devices such as an arc welder in your home. In addition, it is wise to look at your electrical appliance consumption list and see if there is a likelihood that several large motor loads may start at the same time.

Inverter manufacturers have done an excellent job in providing plenty of room for error, so that as long as you are reasonable, there should be no problems with the model selected.

Inverter Temperature Derating

Most inverters' current protection circuitry is temperature compensated; therefore, the maximum size load that an inverter can run changes with the ambient temperature. As the temperature of the internal electronics of the power switching bridge increase, the allowable connected load current/power is reduced.

The graph in Figure 11-7 shows the effect temperature has on a Xantrex model SW series inverter's capacity to operate connected loads. Notice that the inverter

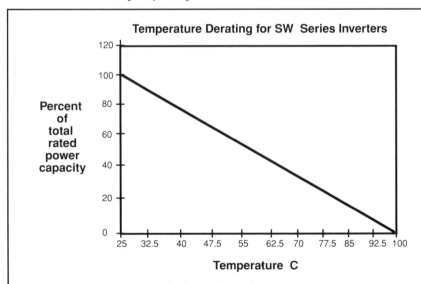

Figure 11-7. Inverters must be derated as ambient temperature increases according to the manufacturers data sheet or this graph. (Xantrex Technology Inc.)

reduces its capacity at temperatures above 77° F (25 °C). The graph also assumes that the inverter is operating at sea level and without any restriction of the airflow around it.

Battery Charging

Applications not requiring battery charging will be able to save money by purchasing an inverter without a battery charger, rather than a combined inverter/charger unit. If you want to install a battery charger after the inverter is installed, a separate battery charger may be purchased such as the TrueCharge™ series from Xantrex Technology Inc.

Summary

Inverters have progressed to such highly advanced products that every cottage and cabin can afford one. This eliminates the need for following in the steps of the renewable energy old timers who had no choice but to use 12 V RV appliances.

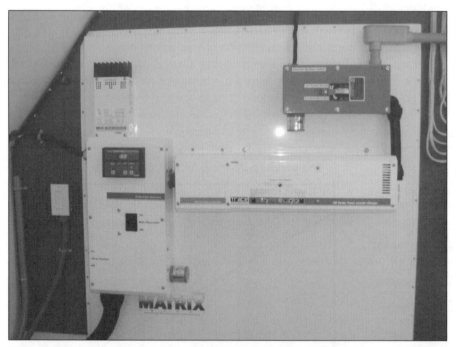

Figure 11-8. A pre-wired panel board contains all the electrical systems required for this mid-sized renewable energy system. In this photo, a 2,500 W inverter is shown at the center, a DC disconnect and circuit breaker are at the lower left, a charge controller is at the upper left, and generator connection is at the upper right

Sun-spiration

"When the local electricity utility was being broken up into separate generation and distribution companies, various energy marketers were going door to door to sign up customers. One arrived at our house. We live so far out in the country that we rarely get door-to-door salespeople, so I was very surprised to see a young man approach me wearing a "Hydro One" I.D. badge and carrying a clipboard with application forms attached. He explained to me that he was here to sign me up to purchase my electricity from them.

I told him that we don't buy electricity - we make it ourselves using solar and wind. I pointed out our solar panels and our wind tower and also showed him that there are no hydro lines running into our home.

He looked very perplexed for a moment, but I could see the wheels turning as he digested this information. I continued to explain our set-up to him, not sure that he understood what I was telling him. Suddenly he asked "no bill?" I assured him that no, we don't receive a bill for our power. A huge grin spread across his face as he repeated over and over again "no bill ... no bill" He thanked me and ran towards the car and driver that were waiting for him at the end of the driveway shouting "no bill, no bill!"

Avoiding a hydro bill wasn't our motivation for going off-the-grid, but I guess it would have worked for him!"

Michelle Mather

Chapter 12 Backup Power Sources

I f there is an Achilles' heel in off-grid renewable energy systems, it has to be the requirement for a backup source of power. Not that there is any problem with equipment reliability. In fact, the opposite is true near our home. Over the last ten years, our neighbors have gone from being more than a bit skeptical to recognizing that while the grid has been down dozens of times our home keeps on ticking.

No, the problem has more to do with the variability of the weather than the equipment. During the dark months of November and early December, we have what appears to be weeks without any sight of the sun or a puff of wind. Once your battery meter tells you its time to charge the batteries, you have to listen.

Bring in the fossil-fueled backup generator. The antithesis of what renewable energy is about. Well, life is always full of

Figure 12-1 Fossil fueled back up generators are the antithesis of what Renewable Energy stands for. Nevertheless, they are an integral part of any off the grid system. (Generac Power System Inc.)

trade-offs, and the generator is one that can't be ignored.

A well-designed and functional off-grid system will operate about 90% of the time without the help of a backup generator. But when the weather decides to pull the clouds over the PV panels' eyes, you have no choice but either shutdown the entire system or start the generator.

A backup generator is designed to perform one primary function. Charge the battery bank to a full state and then shutdown as quickly as possible. Generator power is polluting, noisy and expensive when compared to grid or renewable-source produced electricity. The less generator running time that is required, the better. To limit running time, the inverter's internal battery charger will "load" the generator to a maximum level during the bulk charging step (discussed in Chapter 10). If the battery is being charged "normally", the inverter will switch from the absorption mode to float mode to complete the cycle. If equalization is desired, the battery will complete the high-voltage equalization charge and return to float mode.

Inverters equipped with automatic controls will then signal the generator to stop, completing the charging cycle. Depending on battery and generator capacity as well as depth of discharge, the entire process can take from 5 to 10 hours to complete.

It is common for an off-grid renewable energy system to operate for 9 or 10 months of the year without the generator switching on. When the cold and dark months arrive, the unit had better start and be ready to go.

Generator Types

A backup generator with a reciprocating internal combustion engine is more correctly known as a *genset* (a generator and motor set). They come in many shapes and sizes, and have different fuel supply choices. Before we look at models suitable for off-grid systems let's review what types not to buy.

Small generators, such as the ultra-portable model shown in Figure 12-2, are not suited for off-grid battery charging applications. As a quick rule of thumb, if you can lift the generator, it is probably too small. The Honda model em5000s shown in Figure 12-3 is about the smallest (and least expensive)

Figure 12-2. This small 1,000 W ultra-portable unit from Yamaha is fabulous for operating small power tools or emergency lights, but it is not suited for off-grid battery charging. (www.yamaha.com)

generator I would recommend. If you have an old 4,000 W unit kicking around in the garage, by all means put it to use. Keep in mind that performance, fuel economy and battery charging time will be compromised with undersized models.

Generator Rating

Generators are rated in watts (W) or volt-amps (VA), similar to an inverter. The reason is that motor loads behave differently when connected to an electrical circuit as compared to simple lights and heating units. For most home applications, watts and volt-amps can be used interchangeably.

Inexpensive gensets will have inexpensive generators and support electronics attached. This is not a problem for most applications until big motor-driven devices are activated. Anyone familiar with gensets will know that air compressors, well pumps and furnaces may have a hard time getting started. This often occurs even though the genset

Figure 12-3. The Honda EM5000s or equivalent sized models are the smallest units that should be considered for battery charging applications.

has a higher power rating than the connected load. Motor loads have high starting power requirements that can exceed their normal requirements by three or four times.

Similarly, battery charging can be very hard on smaller, less expensive gensets. This is caused by the unique way in which a battery charging application only consumes power from the very peak of the alternating current waveform. Smaller generators, typically less than 5,000 W, have difficulty providing power in this mode, even though the charging power applied to the battery may be less than half of the generator's rating.

A gas engine driving a generator is a reciprocating device. You will recall from earlier chapters that rotating generators produce an alternating current (AC) with a sine wave pattern, as illustrated in Figure 12-4. Reviewing this waveform, you will see that voltage is represented on the "y" axis, while time is shown on the horizontal "x" axis. Starting from the extreme left of the waveform, the voltage starts out with a zero amplitude and slowly rises until it reaches a peak of 170, at which time the polarity starts to reverse and the voltage drops back to zero before starting on the negative half of the cycle.

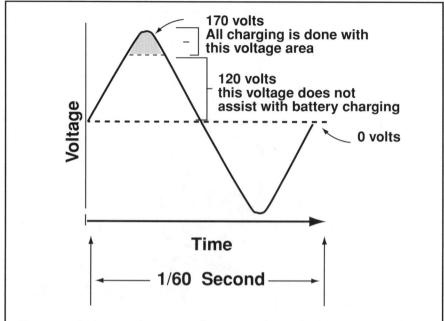

170 volts
All charging is done with
this voltage area

120 volts
this voltage does not
assist with battery charging

0 volts

Voltage

Time

1/60 Second

Figure 12-4. Battery charging applications utilize only the top portion of the generator's sine wave voltage output. Small generators often have difficulty meeting this requirement.

The voltage of the generator source must be higher than the battery for charging to occur. You must also remember that the battery-charging unit inside the inverter contains a transformer that steps up and steps down the voltage. For example, in a 12 V battery bank, the inverter will step this voltage up to 120 V AC to operate the connected loads. Likewise, the inverter will step the generator's 120 V AC down to a DC voltage level sufficient to "push" current into the battery under charge. The waveform in Figure 12-4 shows that the generator's AC voltage starts at zero and climbs to a peak of 170. There will be a period where the generator's "stepped-down" voltage is less than that of the battery bank. No charging current will flow at this time.

Once the generator's voltage exceeds 120 V, the inverter's transformer will step it down by the appropriate ratio, convert the AC to DC and feed the voltage into the battery. The example shown in Figure 12-5 will help to clarify this point. The generator voltage has now risen to 150 V. The applied voltage reaches the inverter or battery charger's transformer where it is stepped down at the ratio determined by the nominal battery bank voltage, which is 10:1 in this example. The transformer output voltage is 15 V AC. This voltage is applied to a rectifier that converts the alternating current to direct current. As the 15 V output is greater than the batteries 12.18 V reading (or 50% discharged), current will flow into the battery.

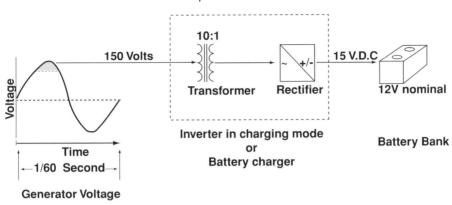

Figure 12-5. It takes a strong, high-quality generator to support battery charging. Weak or inexpensive generators burn excessive amounts of fuel to offset the long run time required to charge a battery.

The "squashing" of the peak of the sine wave identifies gensets with weak or small generators that have insufficient power to support battery charging. Longer charging time and increased fuel consumption result in unnecessary genset wear and increased costs for each watt of electricity stored in the battery when using weak gensets.

It is highly recommend that you purchase a genset with a rating of 7,500 W (120 V/62 A or 240 V/31 A) or higher. Check with the manufacturer to determine

Generator Type	Inverter Type	Typical Maximum Charging Current (Amps)
Honda 800	DR1512	43
Honda 2200	DR1512	57
Homelite 2500	DR1512	11
Honda 3500	DR1512	39
Westerbeke 12.5 kW	DR1512	65

Table 12-1. This table produced by Xantrex Technology Inc. shows the relationship of several generator models charging current when using a 12 V inverter/charger. The Westerbeke 12.5 kW model can charge a battery bank six times faster than a Homelite 2500. This equates to reduced generator running time, wear and fuel consumption.

Figure 12-6. This 10 kW genset from Hawkpower (www.hawkpower.com) operates on propane fuel and is equipped with an auto-start function, allowing the inverter to operate it whenever the batteries require charging.

if the unit is equipped with an electronic voltage regulator module within the generator. High-quality generators often have a peak voltage adjustment that ensures rapid battery charging. Discuss this issue with your generator sales representative. If he or she is not familiar with battery charging applications with a particular model of generator, provide them with a copy of the above text, and ask them to review the issue with the factory. It is pretty tough to return a generator that is too small for your application.

Voltage Selection

The voltage selection of the genset will be determined by the inverter/charger voltage rating. It is possible to operate an off-grid home at either 120 V or 240 V AC or a combination of the two.

As a general rule, if only one inverter is installed, the input voltage will be 120 V. For a two-inverter connection, use 240 V models. See *Chapter 13 - Putting it all Together Safely* for more information on inverter voltage selection and connection.

Fuel Type and Economy

Gensets are available in several fuel choices, including gasoline, natural gas, propane and diesel. The less expensive units tend to be equipped with gasoline engines.

Larger industrial grade models are normally fueled using propane or diesel. The choice depends on several factors such as cost of the unit, proximity to a fuel source and desire for economy and ease of use.

- Gasoline: Everyone is familiar with small gasoline engines that are ubiquitous throughout North America. Cheap and easily fueled, they offer a very short life span when used in off-grid applications. The majority of gasoline engines operate at 3,600 revolutions-per-minute (RPM), which results in rapid wear and high noise levels. Expect a life span of 5 years before a major rebuild is required.

- Natural Gas: Natural gas engines are offered in two varieties: converted gasoline and full-size industrial. The converted engine is really no better than a gasoline engine, except that it offers the advantage of no fuel handling because it is directly connected to the gas supply line. Industrial-sized natural gas engines are of a heavier design and operate more slowly at 1800 RPM. This increases engine life and greatly reduces engine noise. Natural gas is not available in all areas, as the fuel is made available via pipelines.

- Propane: Propane has the same concerns as natural gas with the exception that this is the fuel of choice for off-grid applications; propane may already be the fuel source for other appliances in your home.

- Diesel: For the ultimate in longevity, the diesel engine has the best track record. Diesel units are heavy, long-lasting machines that operate at slow speeds. Fuel economy is highest with a diesel engine, although bringing in a second fuel source will be required.

- Biodiesel: As the name implies, biodiesel is a clean burning, alternative fuel produced from domestic, renewable resources. Biodiesel contains no petroleum, but can be blended at any level with petroleum diesel to create a biodiesel blend (most often at a ratio of 80% petroleum diesel to 20% biodiesel called B20). It can be used in diesel engines with no major (or any) modifications. Biodiesel is simple to use, biodegradable, non-toxic and essentially free of sulfur and aromatics. (See further notes concerning biodiesel fuel below).

The large propane genset shown in Figure 12-6 consumes 1.93 gallons per hour (7.3 litres per hour) when operated at 100% capacity. The same genset is offered in a natural gas fueled model that consumes 144 cubic feet per hour (4,077 lph). An equivalent diesel model from China Diesel Imports (www.chinadiesel.com) requires only 0.78 gallons per hour (3 lph) according to the manufacturer's data sheets. A diesel (or biodiesel) model requires 2.5 times *less* fuel for the same amount of battery charging as a propane model. Over the considerable life span of either model, this translates into a significant savings in operating cost. A gasoline engine's fuel economy is worse.

Another argument against gasoline is the requirement to pay "road tax" when you fill up at the local gas pump. Less expensive colored or "off road" grade gaso-

Figure 12-7. This 8 kW diesel genset from China Diesel Imports is one of the most economical and popular units used in off-grid market. With proper maintenance, these units should provide over 20,000 hours of operating life.

line can be purchased, but may be difficult to locate. Even with the reduction in road taxes, gasoline fuel used in a high-speed (3,600 RPM) gensets will not be as economical as a diesel or low-speed natural gas or propane model.

Generator Noise and Heat

Aside from having to run a genset in the first place, the second most annoying feature of a genset is having to listen to it running. The best way to eliminate this problem is not to operate them at all. The next best plan is to locate the unit a reasonable distance from the house and enclose it in a noise-reducing shed.

A noise-reducing shed is easily fabricated using common building materials or by purchasing a wood-framed tool shed building. The shed should be fabricated with a floating deck floor that does not contact the walls of the building. This construction prevents engine noise and vibration from radiating outside the building. The walls should be packed solidly with rock wool, fiberglass or, best of all, cellulose insulation. The insulation should then be covered with plywood or other finishing material, further deadening sound levels.

All internal combustion engines create an enormous amount of waste heat. A little dryer vent or hole in the wall won't cut it. The unit shown in Figure 12-8 is mounted so that the 18 inch by 18 inch (0.5 x 0.5 m) radiator and fan assembly blow outside the building pointing away from the main house. This arrangement also requires an air intake, which is provided by air passing under the floating deck of the building.

Figure 12-8. Installing a genset in a well-insulated shed with a "floating" floor prevents motor noise and vibration from escaping. Include a large vent fan to remove waste heat, and be sure this vent is directed away from the main house.

The exhaust gas leaves the muffler vertically and passes overhead to a second automobile style muffler further reducing noise and before exiting the building.

Generator Operation

Moving the genset away from the house involves two considerations: a suitable power feed cable of sufficient capacity will have to be run either overhead or in an underground trench (long cable runs increase the cost of installation); the second consideration involves starting the genset. If the unit is equipped for manual starting, it will be necessary to go to the machine shed each time you wish to start and stop the unit. This is no problem on a nice summer day, but it can become a bit trying during winter storms when you need the darn thing the most.

Gensets equipped for automatic operation solve the starting problem. When the AC supply cable is placed in a trench, a second "control" cable is run along side it. This control cable is connected to the genset's automatic control unit, and the other end is connected to a manual start/stop switch inside the house or to the automatic generator controls contained within the inverter. Automatic start functions do not add an appreciable amount to the cost of the genset and greatly improve your relationship with the beast.

Other Considerations

A genset should offer you a long life, particularly if the unit is well maintained during its lifespan. Many underestimate how much television they watch, and I suspect that generator running time estimations follow by a close second. Running time is not a problem, but knowing when to perform periodic maintenance on the unit is. For example, the Lister-Petter diesel engine (the genset model shown in Figure 12-8) requires various servicing functions at 125, 250, 500 and 1,000 hour intervals. Order the unit with a running time meter at a small extra cost.

Oil, air and fuel filters will require changing periodically. Order sufficient spare materials and a service manual for your model. The dealer will be able to recommend a suggested spare parts list.

Have spare engine oil on hand as well as the necessary tools to change filters and drain the engine crankcase. Inquire at your local garage about where used engine oil can be dropped off for recycling. Most garages will be happy to oblige, especially if you deal with them for automotive service. Never pour used motor oil into the ground. One quart (1 liter) of oil can destroy 100,000 quarts of ground water, some of which may feed your well.

To ease winter starting, use synthetic oil rated for operation with your generator model.

Further Notes Concerning Biodiesel Fuel

In recent years, there have been remarkable strides in the "greening" of diesel engines to the extent that many are compliant with California's Clean Air Act requirements. Anyone who thinks that diesel engines are slow, clunky, smelly engines have obviously not been introduced to the new "common rail diesel engine" technologies of the past few years. Witness the Mercedes Benz model E320 series. This car is just as quiet as its gasoline counterpart, and equally as fast in both acceleration and top speed. Cold weather starting problems are also a thing of the past. Not only can you forget about block heaters, but also the whole glow-plug thing is fading out as quickly as carburetors and muscle cars.

As a further advantage, biodiesel fuel (manufactured using soya and other grain plants) may be available in your area. Burning this biodiesel is considered green, as the exhaust components are no worse than those given off by the same plant material rotting in a field. There may be a slight cost penalty for its use, but considering the minimal running time of the genset and the symbiotic relationship this fuel offers with the renewable energy system, it is well worth the price. It might even improve your relationship with your genset.

For more information on this product and where to purchase it, visit www.thesoydailyclub.com or www.aboutbiodiesel.com.

Sun-spiration

"Our son Robert had been invited to a birthday party earlier, so I stayed home to prepare dinner for Jamie and me. He usually arrives home from work around five o'clock, so everything was hot and ready. The only problem; no Jamie.

I called his cell phone to see when he would arrive only to hear it ring and ring. That was odd - usually he carried it with him. I was a little bit concerned, but not overly worried as work had its share of overtime. I decided to wait.

About two hours later, he arrived home. "Where were you and why didn't you call, I was getting worried" I lamented to him. Jamie replied, "Haven't you heard, 50 million people are without power. I had to make sure the building at work was secure and hire a watchman because the alarm system wasn't running!"

I guess I didn't know. Not that it really mattered to us. I popped our meals into the microwave to warm up and turned on the TV to see how everyone else was coping. The renewable energy system in the basement had just earned its keep!"

Karen Wilson

Chapter 13 Putting It All Together Safely

W ell, we have finally made it to the point where we can stop talking about how all the bits and pieces that make up a renewable energy system and start putting it all together, safely. This chapter deals with interconnecting the various system components in a neat and safe arrangement.

If you are not familiar with electrical wiring, conduit and general construction work, don't despair. It is well worth having a look at this section in order to understand what your electrician is talking about during these final steps, *before* "throwing the switch". This section, and the accompanying appendices, also form a reference should your electrician not be familiar with some of the details working with direct current (DC), PV modules, wind turbines and the like.

WARNING!

Grid-interconnected systems are a fairly new phenomenon in North America that may not be allowed in your jurisdiction. Be sure to check with your electrical inspection authority **before** you commit to such a system. Also, be aware that where connection is allowed, there may be no way to be paid for the electricity you produce. Other jurisdictions may be overzealous in their treatment of renewable energy systems, requiring outdoor lockable disconnection switches or expensive "bi-directional" tariff meters. Discuss all of these points with your utility before you commit dollars!

A Word or Two about Safety

Obviously, you want the installation work to be done correctly and safely. Owning and operating a renewable energy system is quite enjoyable, and you can almost forget that you have one at times. Although it is pretty cool stuff, it is not a toy and can cause electrocution or fire hazards if not respected. My first discussion with electrical contractors and inspection people left me bewildered. I clearly remember 10 years ago one person saying, "Why would you want one of those systems? You won't be able to run a toaster". Although I still chuckle at this while eating my morning toast, it goes to show that not everyone is up to speed with the technology.

Owning a renewable energy system is no different than owning and operating a standard electrical power station. In this case, size doesn't matter. You can still die or be seriously injured with battery or inverter power, just as you could from the grid.

Electrical Codes and Regulatory Issues

In North America, electrical installation work is authorized by local electrical safety inspection offices that issue work permits and review the work in accordance with national standards. In the United States, the National Electrical Code (NEC) has been developed for the last 100 years to include almost all aspects of electrical wiring, PV, battery and wind turbine installation. In Canada, the Canadian Electrical Code (CEC) performs the same function as the NEC.

The codes of the United States and Canada form what are known as the Part 1, Installation Codes, and deal with the interconnection and distribution of electricity to industrial, commercial and residential buildings. These codes also deal directly with the internal wiring of your home.

Many people (mistakenly) believe that just because you have your own renewable energy system that the code rules do not apply to them. This is wrong. With few exceptions, the installation of PV, wind and other systems must comply with the requirements of these codes. For example, Article 690 was added to the NEC in 1984 to deal specifically with the installation of PV systems.

In addition to the CEC/NEC code rules, a Part 2 product standard must certify every electrical appliance that operates at 120/240 volts (V) AC. Where safety concerns are present, this scope may be expanded for lower voltage products . Many people are familiar with the Canadian Standards Association in Canada and Underwriters Laboratories in the United States. Working in harmony with the CEC/NEC, these safety agencies are in charge of developing electrical and fire safety standards for appliances. When a manufacturer develops a new product, the design must undergo extensive safety-related tests by these agencies. Products that meet the requirements are eligible to carry a "certification mark", which alerts electrical inspectors that when properly installed, these products are safe.

Legitimate manufacturers will have their products undergo such product testing, and will be eligible to carry the UL, CSA, ETL or other authorized testing laboratory marking. When purchasing and comparing products with one another, look for this marking as a further sign of a safe, quality design.

When you or your electrician are ready to begin wiring, it will be necessary to

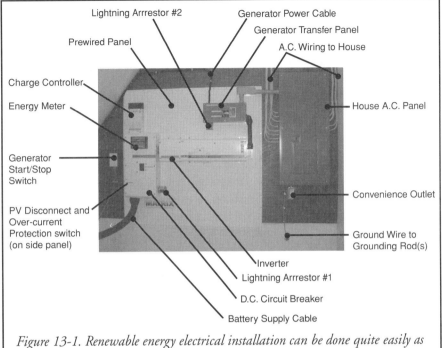

Lightning Arrrestor #2
Generator Power Cable
Generator Transfer Panel
Prewired Panel
A.C. Wiring to House
Charge Controller
Energy Meter
House A.C. Panel
Generator Start/Stop Switch
Convenience Outlet
PV Disconnect and Over-current Protection switch (on side panel)
Ground Wire to Grounding Rod(s)
Inverter
Lightning Arrrestor #1
D.C. Circuit Breaker
Battery Supply Cable

Figure 13-1. Renewable energy electrical installation can be done quite easily as this pre-wired, integrated panel from Xantrex Technology shows. The large gray panel to the right is where the standard 120 V house wiring system begins.

apply for an electrical permit. This permit will authorize you or your electrician to:

- Perform all electrical wiring according to NEC/CEC codes and any local ordinances in effect at the time of installation.

- Install only electrical equipment that is properly certified. Each device must have a UL, CSA or other approved body inspection permit.

- Provide the inspector with copies of wiring plans, proof of certification or other engineering or technical documentation to aid in his or her understanding of the renewable energy system.

- Give the inspector written notification that the work is ready for inspection. You must not cover or hide any part of the wiring work, including backfilling of trenches until the inspector has approved the work.

Your electrical inspector is not working against you. If they are asking a lot of questions it is to understand what you are doing. Renewable energy systems are not as yet considered mainstream technology, and some inspectors may not be familiar with the specifics. On the other hand, your inspector will know an awful lot about wiring and installation details, and most professionals will be more than happy to assist with guidance and pointers.

Electrical inspectors will review the installation work and check for certification marks on the various appliances. Some renewable energy system components may not have UL, CSA or other test certification markings. Electrical code rules require that products *must* have such markings. Uncertified products may not be acceptable to your inspector. It is highly recommended that you review the products before you purchase them to ensure proper compliance. If a product is required and the manufacturer has not had the product tested, discuss this with your inspector before you purchase it. A certified product may be available, albeit at a higher cost. Alternatively, field inspection on-site may be allowed for an additional fee.

Caution!

As code rules are updated on a regular basis and may have subtle differences from one locale to another, use the information in this chapter as a guide. Discuss the details with your electrician and inspector before proceeding with installation work.

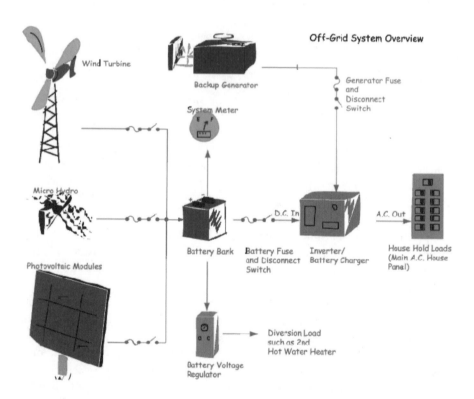

Figure 13-2. This off-grid system overview shows the placement and basic interconnection of each component. Grid-interconnected systems may or may not have batteries or backup generators as noted in Figure 6-5.

What Goes Where?

Referring back to the previous chapters, we have dealt with each component as a separate piece of the pie. We are now ready to begin the wiring installation planning.

As the wiring overview in Figure 13-2 outlines, wiring interconnects each component of the renewable energy system. PV arrays always operate using direct current (DC) and may be connected directly to the battery bank for storage. Wind and hydro turbines may operate using direct or alternating current (AC), although in the case of the latter, a conversion unit known as a *rectifier bridge* will always convert the voltage to DC. The feed from the battery to the energy meter, charge control, diversion load and inverter input are likewise accomplished using DC.

The AC connections in the system are between the generator, inverter and house supply panel.

The distinction between DC and AC wiring is very profound. Wire, connectors, fuses and switches are generally not interchangeable. Because the current on the DC side is very high (owing to the lower voltage), wire size tends to be much larger than on the AC side. As a result, we will review each wiring configuration separately.

DC Wiring Overview

Energy stored in a battery bank is at a low, DC voltage. When energy is required for our homes, the DC voltage is stepped up to either 120 V or 240 V and converted to AC by the inverter. If we assume that the inverter will supply a maximum house load of 1,500 watts (W) it should be apparent that 1,500 W, plus an allowance for inefficiencies, would flow out of the battery bank. The wattage, voltage and current relationship is:

1,500 W load ÷ 120 V house supply = 12.5 Amps (A) current flow

And on the low voltage side of the inverter:

1,500 W load ÷ 12 V battery supply = 125 A current flow

A quick rule of thumb is to remember that the low-voltage side of the inverter current is 10, 5 and 2.5 times greater than the 120 V AC side when dealing with a 12 V, 24 V or 48 V battery, respectively.

It stands to reason that since the wattage is the product of the voltage and current, lowering the voltage will cause a corresponding rise in current and vice versa. It requires a wire the size of a Polish sausage to carry 125 A of current in a 12 V circuit yielding 1,500 W. The same 1,500 W can be supplied through a light-duty extension cord if the voltage is cranked up to 120 V.

This is the one reason why battery voltage selection is based on power and energy requirements. The larger the power requirement and the lower the voltage, the larger the wire must be; conversely, the higher the power and voltage of the circuit, the smaller the wire. Copper wire is expensive and hard to work with, offering us plenty of reason to stick with higher battery voltages where possible. This is

Figure 13-3. Direct current wiring circuits, such as the battery interconnection cables shown here, are expensive and difficult to work with. Keep wiring runs as short as possible and purchase pre-manufactured cable such as these, whenever possible.

also the reason that large energy consumers in the home such as electric stoves, furnaces, central air conditioning and dryers are always rated 240 V.

Low voltage electricity is also difficult to transmit at great distances. As we increase the current to compensate for the lower voltage, more of our power is lost due to the resistance of the wire. Again, the only way to offset this problem is raise the voltage, decrease the transmission distance or increase the size of the wire. Often it is necessary to do all three. In short, keep DC wiring runs as short as possible, increase battery voltage where practical and try to keep load current to a reasonable level. Appendix 10 contains tables that relate the system voltage and load current to wire size and provides a maximum one-way cable length. This chart is calculated to allow a 1% electrical voltage loss. If you can tolerate a higher level of loss, then each of the applicable distances may be doubled, tripled, etc.

For example, refer to Appendix 10, Table "B" for 24 V systems. If we have a PV array that is delivering 40 A of current, and we need to run the cable a total of 24 ft (7 m), the chart indicates that we must use a #0 size of wire (some trades also refer to this as "#1/0", while "#00" is referred to as "#2/0"). This stuff is pretty big (about the diameter of a pencil), hard to work with and fairly costly. So let's look at some of our options:

1. Do nothing. Use the #0 wire and call it a day. There is nothing really *wrong* with using #0 wire; bigger wire is simply more expensive, harder to work with and difficult to interconnect.

2. Increase the system voltage. Increasing the system voltage to 48 V will allow the cable size to be decreased to #6 size (see Table "C"). (Remember, doubling voltage halves the current to 20 Amps.)

3. Allow an increase in voltage loss. Leave the system voltage at 24 V, but allow the system voltage drop to increase to 2%. Use the 24 V data table and reduce wire size as desired. For example, using a #6 wire will increase losses to 4%, which is determined by

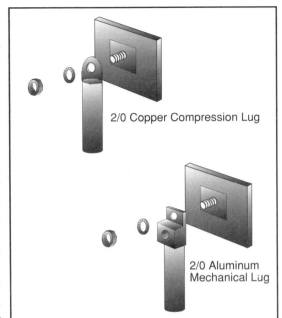

2/0 Copper Compression Lug

2/0 Aluminum Mechanical Lug

Figure 13-4. DC wiring requires terminations that are clean and electrically reliable. The copper compression lug (shown above left) is a pre-manufactured cable. The aluminum lug (shown above right) can be assembled on-site without the need for special tools.

taking the ratio of their 1% loss distances, (24÷6 feet = 4%). Be sure to check the table in Appendix 11 to be sure the selected wire size can carry the required current. Number 6 wire is rated for a maximum of 75 Amps. Don't go overboard with losses to increase wire run distances. Renewable energy is expensive, with the losses reducing power at the batteries or inverter.

4. Run two sets of smaller wire. Suppose that our example has us connecting a PV array to a battery bank. It is possible to "split" the wiring of the array in two, reducing the current in each set to 20 A. If each set of arrays were connected with #4 wire, we can still maintain our voltage drop at 1%, and use smaller wire, which is easier to handle. The two arrays would be connected in parallel back to the battery bank, combining to produce our 40 A supply.

The low voltage side of an inverter requires an enormous amount of current as we learned in the example provided earlier. Therefore, the larger the battery cables connected to the inverter the better. Undersized cables result in additional stress on the inverter, lower efficiency, reduced surge power (required to start motor operated loads) and lower peak output voltage. Don't use cables that are too small and degrade the efficiency that you have worked so hard to achieve.

In addition, keep the cable runs as short as possible. If necessary, rearrange your electrical panel to reduce the distance between batteries and the inverter. The lower the DC system voltage, the shorter the cable needs to be. If long cables are required, either oversize them substantially, or switch to a higher system voltage as discussed above.

Although large cable may seem expensive, spending an additional few dollars to ensure proper performance of your system will be well worth the investment. Xantrex Technology recommends for their inverter systems to tape the positive and negative wires together forming a parallel set of leads. This reduces the inductance of the wires resulting in better inverter performance.

Cable Size Required	Rating in Conduit	Maximum Breaker Size	Wire Rating in Air	Maximum Fuse Size
#2 AWG	115 amps	125 amps	170 amps	175 amps
00 AWG	175 amps	175 amps	265 amps	300 amps
0000 AWG	250 amps	250 amps	360 amps	400 amps

Table 13-1. Battery and inverter cable sizing chart. Note that wires run inside a conduit have a much lower rating than those exposed to air due to heat losses. Battery-to-inverter cable runs should not exceed 10 feet in one direction. It is recommended that you use #0000 AWG size wire for all inverter runs, regardless of length.

AC/DC Disconnection and Over-Current Protection

For safety reasons, and to comply with local and national electrical codes, it is necessary to provide over-current protection and a disconnection means for all sources of voltage in the ungrounded conductor. This includes the connection between the PV array, wind turbine, hydro turbine and batteries, as well as between the batteries and inverter. Most AC sources are already provided with a protection and disconnection device as an integral part of the house wiring. Generator and inverter units are generally provided with their own internal certified fuse or circuit breaker device.

Standard AC rated circuit breakers and fuses will not work with DC circuits, and such an installation should never be attempted. Fuses, such as those shown in Figure 13-5 (top), may be safely utilized as they are rated for breaking a DC source. Circuit breakers, such as the battery to inverter disconnection and over-current module shown in Figure 13-5 (left), must carry this same DC rating.

Fuses and circuit breakers are similar to safety valves. When the flow of current through a conductor or appliance exceeds a specified rating, the fuse will "blow",

Figure 13-5 a, b. A "T" series, DC fuse is shown above and a direct current disconnection and over-current protection circuit breaker and mounting box is shown at left. (Xantrex Technology Inc.)

opening the connection path. A circuit breaker works in the same manner, except that it may be used as a temporary servicing switch and can also be reset after a trip condition.

Each circuit path will have a maximum current path based on worst-case conditions. This current level must be calculated by reading the manufacturer's data sheets for the product. This worst-case current should be provided with a safety factor of 25%. Therefore, when sizing a cable for a run between a PV array and batteries or other appliance, the cable run distance, system voltage, worst-case current and wire current carrying capacity have to be considered. In addition, if the cable is contained in conduit that is exposed to the summer sun, the insulation temperature rating must also be considered. For example:

1. A PV array outputs a maximum of 30 A under all conditions and the one way wiring distance is 40 ft (12 m). The system voltage is 24 Vdc.
2. From the chart in Appendix 10, we see that a #00 (#2/0) wire is required to carry this current with a maximum loss of 1%.
3. From the chart in Appendix 11, we see that a #00 wire is capable of carrying a maximum current of 195 A.
4. A safety factor of 25% is added to our maximum PV array current (30 A x 1.25 = 37.5 A.
5. The worst-case current calculated in #4 above is compared to the maximum current rating of the wire calculated in #3. If the worst-case current is less than the desired wire size capacity, it is okay to use.
6. The circuit breaker or fuse size must be equal to or less than the maximum rating of the wire capacity, as defined in the NEC/CEC. Note that electrical cables run in conduits or raceways must be derated due to heating effects.

7. The ambient temperature of the environment and other conditions affecting wire insulation temperature ratings are discussed in Chapter 2 of the NEC and Chapter 12 of the CEC.

These notes show you the complexity of wire selection, placement and wire type. If you are not familiar with the above, it is recommended that you purchase a copy of the latest NEC/CEC and review the corresponding issues. Alternatively, you may discuss these items with your electrical inspector at the planning stages. To purchase the NEC contact:

National Fire Protection Association
Batterymarch Park,
Quincy, Massachusetts
02269
www.nfpa.org

To purchase the CEC contact:
Canadian Standards Association
178 Rexdale Blvd.
Rexdale, Ontario
M9W 1R3
www.csa.ca

Battery Cables

According to the NEC, battery cables must not be fabricated with arc welding wire or other non-approved cable. Standard building-grade wire must be used. The CEC has no such restriction.

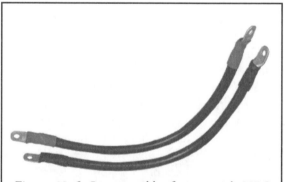

Wiring Color Codes:

Wiring color codes are an important part of keeping the system interconnection straight when installing, troubleshooting or upgrading the system at a later date. The standard color schemes used are:

Figure 13-6. Battery cables for use with NEC approved systems must be approved building wire. The CEC does not have this restriction.

Bare copper, green, or green with a yellow stripe

Used to bond exposed bare metal of PV modules, frames, inverter chassis, control cabinets and circuit breakers to a common ground connection (discussed later).

The ground wire does not carry any electrical current except during times of electrical fault.

White, natural gray insulation

This cable wire may also be any color at all, other than green, provided the ends of the cable are wrapped with colored tape to clearly identify it as white. This wire carries current and is normally the negative conductor of the battery, PV, wind or hydro system. The white wire is also bonded to the system ground connection, as detailed in Figure 13-7 and as discussed later.

Red or other color

Convention requires that the red conductor of a two-wire system is the positive or ungrounded conductor of the electrical system. This requires the ungrounded conductor to be any color except green or white. The majority of DC systems are based on this standard color code.

The AC side of the circuit utilizes a similar approach to color coding, except that the ungrounded conductor(s) are generally black and red (for the second wire).

System Grounding

Grounding provides a method of safely dissipating electrical energy in a fault condition. Yes, that third pin you cut from your extension cord really does do something. The "something" is to provide a path for electrical energy to flow when the insulation system within an inverter or cable covering fails.

Imagine a teakettle for a moment. Two wires from the house supply enter the teakettle, plus a ground wire. During normal operation, the electricity flows from the house electrical panel via the ungrounded "hot" conductor to the kettle. Current flows through the heater element and back to the panel via the "neutral" conductor that is grounded. A separate ground wire connects the metal housing of the kettle to a large conductive stake driven into the earth, just outside the house.

If the insulation or hot wire were to be damaged inside the kettle, it is possible that it may touch the metal chassis. Because the metal chassis is bonded to ground, electrical energy will travel from the chassis, through the ground wire to the conductive stake. This flow of current would be unrestricted due to the bypassing of the element, overheating the electrical wires. If it were not for the circuit breaker or fuse opening during this excessive current flow, a fire may start.

Electrical energy has an affinity for a grounded or "zero potential" object and will do what ever it takes to get to there. If there were no ground connection on the kettle's chassis, electricity would simply stay there until an opportunity arose to jump to ground. If you happen to touch a kettle in such a state and were to simultaneously touch the sink or be standing on a wet surface, the electricity would find its path through your body. This is not a good situation.

In a similar manner to our teakettle, the entire exposed metal and chassis of our renewable energy system are bonded to a *common* ground point as illustrated in Figure 13-7. The white or negative wire of the system is also bonded to this point, saving us from adding a second set of fuses and disconnection device.

The size of the ground wire is determined by the NEC/CEC, and is based in part on the size of the main over-current protection device rating. This is noted in Table 13-2.

Lightning Protection

If you flip back to Figure 13-1, you will notice two lightning arrestors attached to the main DC disconnection panel and the generator transfer panel. These devices contain an electronic gizmo known as a Metal Oxide Varistor (MOV), which connects between the DC +/- conductors and ground as well as the AC hot/neutral conductors and ground. A lightning arrestor is about the cheapest piece of insurance you can purchase to protect your power system. Connect one on every cable run that strings across your property. The system shown in Figure 13-1 has a roof mounted PV panel and a generator located in a remote building. Cables running here and there can attract lightning on its way to ground potential. The arrestor "clamps" any stray voltage and passes it safely to the grounding conductor. Sprinkle one on each of your wind turbine, hydro generator or on AC power lines interconnecting out buildings and barns.

Interconnecting the Parts

A mechanical layout of your desired installation will help determine material required and assist in visualizing the layout of the power station. It is also important to determine what functions you require from the overall design. Some configurations that can be "wired" into a renewable energy system are:
- off-grid system with a generator backup
- on-grid system without battery backup
- on-grid system with battery back
- on-grid system with battery back and generator emergency supply

The number of possible configurations is extensive and we cannot cover every installation type within these pages. Fortunately, the inverter and equipment manufacturers' installation manuals will be able to assist with reconfiguring for custom design requirements. We will focus our attention, covering off-grid design with generator backup, as this tends to be the most complex installation.

Referring back to Figure 13-1, let's now review the specific issues related to the installation of each identified component.

PV Array

PV arrays may be mounted on a house roof, ground mount or sun tracking system. The interconnection of the modules is similar with only one difference between various installation requirements. The NEC requires that every roof-mounted PV array be equipped with a device known as a *Ground Fault Interrupter* (GFI). The GFI unit shown in Figure 13-9 automatically disconnects the PV array in the event

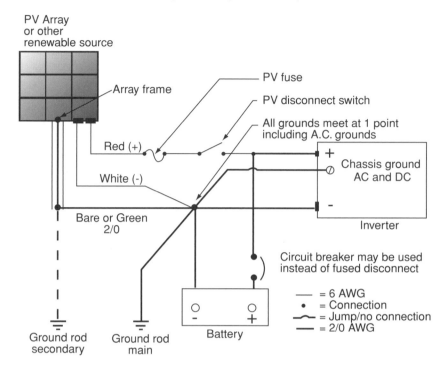

Figure 13-7. This view of a PV array, battery and inverter DC connection shows the over-current protection fuses and disconnect switch in the ungrounded (+) conductor. The negative conductor (white) is bonded to ground, and therefore requires no disconnection and over-current protection device. The ground wire (bare or green) connects at one point to a ground rod. A secondary ground rod may be used where the array or generator is a long distance from the main rod. Note the use of an interconnecting cable between both rods.

Size of Largest Over-Current Device	Minimum Size of Ground Conductor
Up to 60 A	#10 AWG
100 A	#8 AWG
200 A	#6 AWG
300 A	#4 AWG
400 A	#3 AWG

Table 13-2. This table shows the minimum size of the grounding conductor based on the largest over-current device supplying the DC side of the inverter.

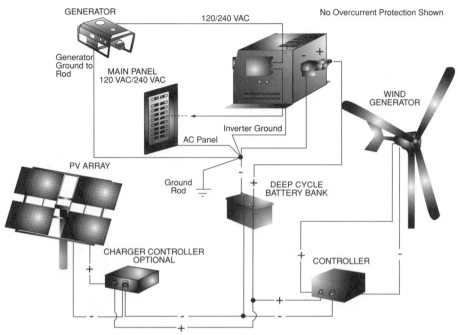

GENERATOR
120/240 VAC
No Overcurrent Protection Shown
Generator Ground to Rod
MAIN PANEL 120 VAC/240 VAC
WIND GENERATOR
Inverter Ground
AC Panel
PV ARRAY
Ground Rod
DEEP CYCLE BATTERY BANK
CHARGER CONTROLLER OPTIONAL
CONTROLLER

Figure 13-8. This simplified schematic shows a typical off-grid system DC and AC wiring arrangement. The charge controllers in this system are wired as series controllers.

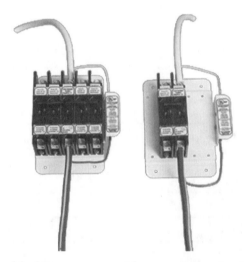

Figure 13-9. A ground fault interrupter provides protection against internal PV module faults that may cause overheating and fire. They are required when the PV array is roof mounted on a dwelling building.

of over-current or insulation fault that may cause overheating and a possible fire. A GFI is not required on ground, pole or tracker mounts.

The first step in wiring your PV array is to determine what battery voltage you will be operating at. We discussed in *Chapter 6 – Photovoltaic Electricity Production* how a standard module is wired for 12 V output. Chapter 6 also covered the steps required to increase PV array voltage by wiring modules in series as well as in parallel connection for increasing current flow. Figure 13-10 shows a group of four 12 V modules interconnected together in series, forming a 48 V string. Each module is provided with a weatherproof junction box and knockout holes to place liquid tight strain relief bushings. These bushing press into the hole in the junction box and are held in place by a retaining nut. The flexible cable is then passed between junction boxes and the series or parallel interconnection is completed as described in Chapter 6.

Figure 13-10. A PV array detail showing four 12 V modules interconnected in series, forming a 48 V string. The output of each of four strings feeds into the combiner box shown at the center.

Be careful to check for proper wire gauge and type to be sure it is suited for outdoor, wet installation. Review the wire choice with your electrical inspector or the NEC/CEC code rulebook.

The output from a grouping of several modules then meets at a combiner box, which is shown in detail in Figures 13-11 and 13-12. The combiner box in Figure 13-11 is mounted on the rear of a sun tracker unit. In this array, a total of sixteen 12 V modules are connected in eight sets of two pairs. Each module is rated 75 W. The two modules are wired in series forming a 24 V set. Each of the eight sets' wires are then directed to the combiner box where all of the positive and negative wires are connected together in parallel. The parallel connection is completed using split bolts or terminal strips suitable to connect all of the appropriately sized wires to the main house supply cable.

Digressing a moment, lets look at how this works when fully connected:

a) *16 modules x 75 W per module = 1,200 W for the array*

b) *2 x 12 V modules wired in series = 24 V*

c) *Current from array = 1,200 W ÷ 24 V = 50 A*

If we had simply left the array wired at 12 V and paralleled the 16 modules we would have had twice the current (100 A). Likewise, if the array were to be wired at 48 V, the current would be 25 A. Power in watts remains the same regardless of how we interconnect the modules.

The combiner box shown in Figure 13-11b contains the GFI unit discussed earlier.

Wind and Hydro Turbine Connection

Connection of a wind or hydro turbine is done essentially the same way as the PV array. Each model will provide a connection point or suggest a wiring interconnection method in the associated installation manual. For example, Bergey WindPower recommends strapping a weatherproof junction box near the top of the tower. This junction box is drilled to allow the feed/connection wires from the turbine to enter the box. A weatherproof fitting is used to provide strain relief for the wires and keep the weather out.

A feed cable is then run down a tower leg to a second servicing junction box. The feed cable running the length of the tower (and to the house) should be installed in either a PVC plastic conduit or metal conduit secured to a tower leg with non-corrosive metal clamps. As an equally acceptable alternative, investigate a flexible cable/conduit combination known as Teck cable. This material is more expensive than traditional conduit but installs in seconds and as is just as strong.

Hydro turbines can be installed using PVC conduit or Teck cable. Both types of installation are weatherproof and will work well in this application. Note that conduit is not suited for movement once installed. If the turbine will be moved for servicing or repair, Teck cable offers the best choice.

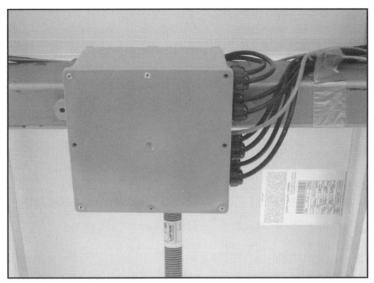

Figure 13-11 a

Figure 13-11b. These combiner boxes are fabricated using weatherproof junction boxes and strain relief. The feed wires from the arrays feed into the boxes and are parallel with the main house feed cable. Figure 13-11 b, this combiner box contains the GFI required for roof-mounted PV arrays.

Renewable Source to Battery Feed Cable

The electricity from the PV array (wind or hydro turbine) must be routed to the battery room. Cables may be wired overhead or underground. Underground cable connections tend to be the most common due to simplicity and lower installation cost. The NEC/CEC have provisions for both direct burial cable and cable protected by conduit. In either case, a trench is dug, typically 18 inches (0.5 m) deep between the source and the battery room or house. Where the cable exits the ground to enter the house or travel up the tracker support tube, it is necessary to use a length of conduit to protect the wire from damage. The cable is placed in the trench and covered with 6 inches (15 cm) of soil. At this point, a marking tape is placed into the trench. The intent of the tape is to warn anyone digging in the area that a buried cable lies below. The trench is then filled completely filled in.

Over Current Protection and Disconnect Devices

For the energy source to be connected to the inverter (for grid-interconnected systems) or batteries (for off-grid systems), the NEC/CEC requires that an over-current protection and disconnection device be installed. This may be done in one of two ways: you may purchase individual fused disconnect switches for each source, as shown in Figure 13-12, or have auxiliary circuit breakers added to the main battery/inverter disconnect box. The latter was chosen for the system outlined in Figure 13-1.

Battery Wiring

As we learned in *Chapter 9 – Battery Selection and Design*, wiring batteries in series increases the voltage (Figure 13-13a), and connecting them in parallel increases the capacity (Figure 13-13b). Battery manufacturers offer may voltages and capacity ratings to suit your specific requirements. The most important considerations for installation are voltage, capacity in amphours (Ah) and distance from the inverter.

As the size of the battery bank physically increases it becomes more difficult to keep the wire lengths feeding the inverter as short as discussed in Table 13-1. Plan

Figure 13-12. Individual fused disconnection boxes, such as the one shown here, may be used for each renewable source of energy, or they may be combined within the main disconnect chassis.

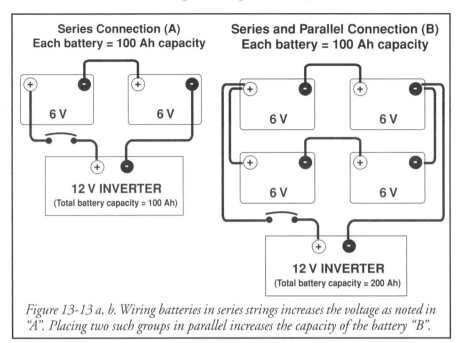

Figure 13-13 a, b. Wiring batteries in series strings increases the voltage as noted in "A". Placing two such groups in parallel increases the capacity of the battery "B".

the battery room layout using graph paper and cut-outs of the selected battery model to determine the best physical layout, paying close attention to the cable connections and run lengths. Be sure to allow for battery disconnection and over-current protection within the wiring layout. Remember to "play" with different battery sizes, shapes and voltage arrangements based on your system configuration (see Table 9-2 in Chapter 9).

To better visualize this concept, the battery bank shown in Figure 13-14a is comprised of two shelves of eight batteries rated at 6 V each. We discussed in Chapter 9 how each battery's cell generates a nominal 2 V DC. In the photograph, you can see three cells for each battery; three cells x 2 V per cell = 6 V. If you look carefully, you can see the battery cables connected from one battery to the next creating the series chain. The drawing in Figure 13-14b details this connection in a schematic view. The battery bank on the upper shelf is also wired in series forming a 48 V bank. The two banks are then placed in parallel in the same manner as outlined in Figure 13-14b.

The battery capacity and physical size of this installation are so large that the length from one end of the battery bank exceeds the manufacturer's recommended cable length for inverter operation. To circumvent this problem, interconnection wires were bumped up to the largest size possible (#4/0) to try and mitigate this problem. Sometimes there is only so much you can do!

Figure 13-14a. This large battery bank comprises a series string of two shelves of eight 6 V batteries generating 48 V. The second identical string on top is wired in parallel with the first, increasing capacity.

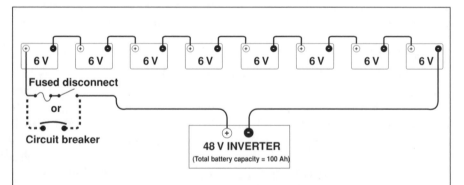

Figure 13-14b. This schematic details one shelf of batteries shown in Figure 13-14a. The second shelf of batteries would be wired in the same series string manner. The first and last batteries would then be parallel connected between shelves in the same manner outlined in Figure 13-13b.

Battery Voltage Regulation

Battery voltage is regulated using a charge controller, which was discussed in *Chapter 10 – DC Voltage Regulation*. The charge controller may be connected as either a series regulator or shunt regulator. The series regulator is the simplest arrangement, simply turning on and off the PV input to the batteries based on their state of charge. If your system includes a wind or hydro turbine, it's almost always necessary to maintain a load on the generator, requiring a shunt or diversion load arrangement.

Series Regulator

Series regulation is by far the simplest and least expensive system to install. The downside with series regulation is excess energy is wasted when the unit is maintaining voltage. Morningstar Corporation indicates that series regulation adds less than 5% to the cost of a mid- to large-sized PV array and battery system.

Figure 13-15 shows the connection details for a Xantrex Technology C-40 model charge controller. The connection details can't get much simpler than this. Note that circuit breakers (or fused disconnect switches) must be used in each leg of an ungrounded wire (positive lead) fed by a source of voltage.

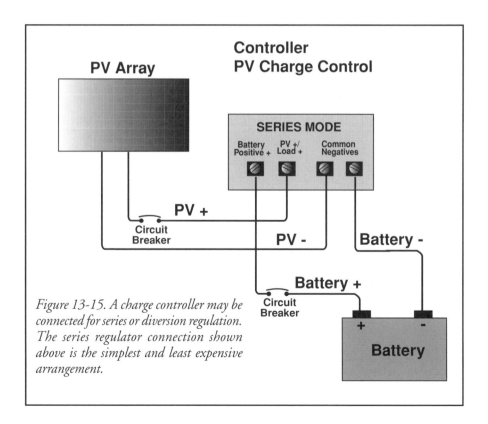

Figure 13-15. A charge controller may be connected for series or diversion regulation. The series regulator connection shown above is the simplest and least expensive arrangement.

Shunt or Diversion Regulator

When a shunt regulator is doing its thing, it really makes me smile. Picture this: a lovely summer morning, the batteries are bubbling away, fully charged. The PV/wind/hydro source power is no longer required. The charge controller starts sending this "waste" power into the hot water tank, reducing our propane demand. This is synergy at its best.

Admittedly, is takes a little more equipment and space to perform the installation. Fortunately, the equipment is not very costly or complicated to get working. But is it for you? This depends on a simple factor, estimating if there will be excess energy to make the installation worthwhile.

Calculating the amount of excess energy is not a simple task until you have lived with the system a year or so. The reason is due to the variability of the renewable energy source and the changing energy requirements in the home. For most homeowners, a PV-based system will increase its output by 200% or more from winter to summer. If you spend a great deal of time outside in the summer months, electrical loads will be reduced.

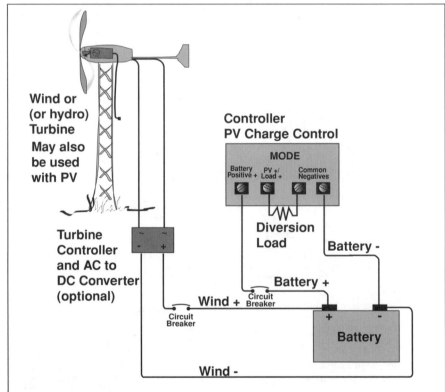

Figure 13-16. A shunt or diversion regulator requires connecting the charge controller to a diversion load. This load can be a hot water heater or electric air-heating element.

One option is to install the charge controller in a series configuration to start and plan for an upgrade later. The major issue with respect to the upgrade is pre-installing the wiring from the charge controller to the water heater tank. The next issue is to leave room for an electric water heater if an upgrade is completed at some future date.

Monitoring when the batteries become fully charged during the summer months will determine the viability of such a system. For example, if your batteries are charged by 11:00 AM on sunny days and your panels provide full output power until 6:00 PM, with a 1,500 W array, we can calculate the "excess" power:

7 hours of "excess" production x 1,500 W PV array = 10,500 W energy

In this example, that's equal to a mid-sized electric water heater (3 kW) operating between 3 to 4 hours when grid connected. Not bad, that should produce a full day's hot water for the average conserver family. At the very least, it reduces hot water heating fuel from a non-renewable resource.

So what does it take to make this happen? Review Chapter 10 for details on plumbing and electric heating element swapping. The standard 120 V/240 V elements **must** be replaced on 12 V or 24 V systems. The 48 V system may be able to use the existing heating elements, although at a lower wattage, in accordance with Table 13-3. (Water heaters of differing wattage may have the wattage ratio adjusted accordingly. For example, a 1,500 W, 120 V element will produce half the wattage of that shown in the chart). When using existing heating elements in a 120 V or 240 V water heater, make sure the internal thermostats and over-temperature protection devices are bypassed, either by removing them from the circuit or installing a suitable heavy gauge jumper wire.

It is possible to use air-heating elements as a diversion load, such as the model shown in Figure 10-6. Air heating tends to be wasteful as the majority of the heat is generated in the hot summer months.

System Voltage	Power (Watts)	Current (Amperage)
120 Vac	3,000	25
15 Vdc (12 V nominal)	48	3.2
30 Vdc (24 V nominal)	190	6.3
60 (48 V nominal)	750	12.5

Table 13-3. This table shows the effects of using a standard 120 V water heater on various battery voltage systems. The system voltage noted is the maximum diversion voltage with batteries fully charged. As you can see from the chart, 12 V and 24 V systems do not have sufficient load to work with 120 V heating elements.

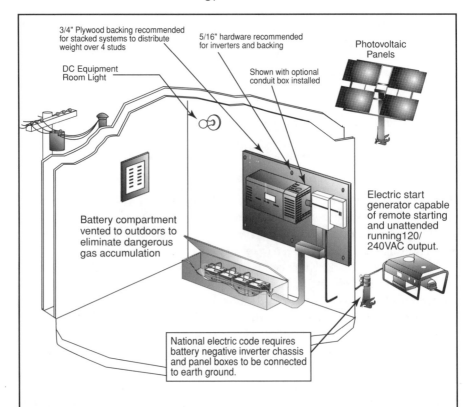

3/4" Plywood backing recommended for stacked systems to distribute weight over 4 studs

5/16" hardware recommended for inverters and backing

Photovoltaic Panels

DC Equipment Room Light

Shown with optional conduit box installed

Battery compartment vented to outdoors to eliminate dangerous gas accumulation

Electric start generator capable of remote starting and unattended running120/ 240VAC output.

National electric code requires battery negative inverter chassis and panel boxes to be connected to earth ground.

INSTALLATION NOTES

DC disconnect and overcurrent protection for inverters, 15 amp branch circuit rated DC breaker for equipment room light and other small DC loads. All system grounding terminates at this unit. System meter shunts are accomodated . Non- stacked installations can accomodate up to 100 amps of DC load circuits.

Solar charge control with disconnects. Breakers DC rated up to 125V and 25KAIC GFI required for roof mounted arrays.

Conduit required for all high current and high voltage wiring unless otherwise adequately protected.

Note: Conduit runs under 24" in length are allowed up to 80% fill and do not require derating for more than 3 conductors. Systems capable of 4KW continuous operation per inverter for periods in excess of 3 hours are subject to additional ampacity derating.

DC Equipment ground does not need to be in same conduit as positive and negative conductors but must be afforded equivalent physical protection.

Figure 13-18. This overview drawing details many of the issues related to the DC connection of the batteries, inverter, charge controller, disconnect and over-current protection system. Consult with the NEC/CEC and your electrical inspector to determine the details specific to your installation. (Drawing based on Xantrex Technology Inc. installation designs).

Directing the excess heat to a hot tub or spa is another good place to "dump" excess power. Use caution when connecting the diversion elements in this application. Both Underwriters Laboratories and Canadian Standards Association safety standards require the use of current collectors and GFI protection in spa systems (note 1).

Note 1: UL Standard UL 1795 and CSA standard C22.2 #218.1 require the use of current collectors to prevent shock hazard in the event of heater failure. 120 V spa heating elements may be purchased from your local pool and spa supply store. Consult with the applicable standards before connecting the diversion controller in this situation.

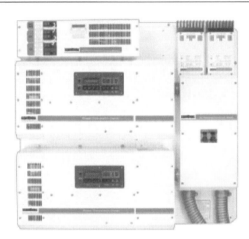

Figure 13-17. This pre-wired electrical system is compact, neat and properly installed, which reduces problems with sometimes-wary electrical inspectors. (Xantrex Technology Inc.)

The Inverter
DC Input Connection

The inverter is the heart and brains of the renewable energy system. Inverters are also pretty darned heavy, and if your system is going to be designed to operate 240 V house loads, it may be necessary to have two inverters "stacked" together to generate this voltage. Figure 13-17 shows a pre-wired power panel from Xantrex Technology that contains:

- two 4,000 W sine wave inverters
- two series wired charge controllers
- dual 250 A over-current disconnect units
- AC wiring chassis for generator and house panel connection

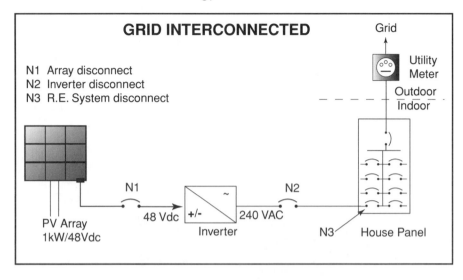

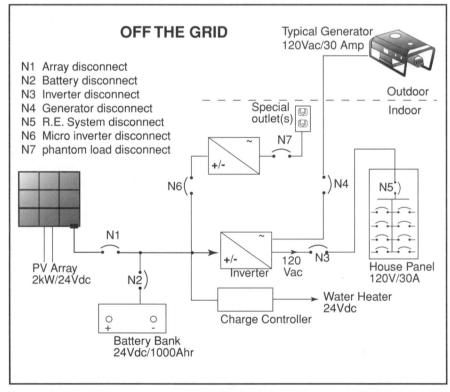

Figure 13-19 a, b. Interconnection to the electrical grid will require a single line diagram schematic detailing the system design. Your electrical inspector and/or electrician can help you configure the safety elements of the systems to meet your particular needs.

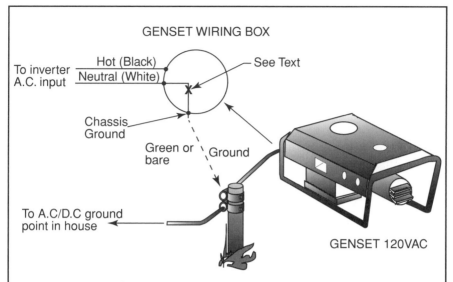

GENSET WIRING BOX

To inverter
A.C. input
Hot (Black)
Neutral (White)
See Text

Chassis
Ground

Green or
bare
Ground

To A.C/D.C ground
point in house

GENSET 120VAC

Figure 13-20. In order to ensure only one ground connection point in the system, the neutral to ground bond inside the generator will have to be removed as noted above.

Purchasing your system pre-wired like the one shown in Figure 13-17 is a wise decision. As the panels are assembled at the factory and approved to applicable safety standards, there is no problem satisfying your electrical inspector. In addition, costly errors in wiring are eliminated.

If you wish to complete your own wiring, Figure 13-18 details some of the areas of concern in making the connection. Note that this schematic drawing is in no way a full interpretation of the NEC/CEC code rules, but it does provide you with a basis for planning with your electrical inspector.

AC Output Connection

So far we have categorized renewable energy systems as being either on or off the electrical grid. While this generalization is accurate, what it does not reflect are the literally dozens of ways the systems can be wired and interconnected together. Some of the more common system configurations are:

- off-grid with no genset backup power source
- off-grid with genset power source
- on-grid, no battery storage
- on-grid, battery storage for short term outage protection
- on-grid, battery storage and genset for long term outage protection

(Note that each system may be connected for either 120 V AC only or 120/240 V AC operation, typically using two inverters.)

This basic combination of systems produces ten different wiring methods with many more designs being possible. It will be necessary to determine which system configuration suits your requirements and then consult with the various equipment manuals to determine proper wiring configurations. For the purposes of this Handbook, we will assume a common off-grid configuration and a basic grid-interconnection design without battery storage, which is typical of most applications.

It does not matter which system you wish to install, your electrical inspector will want to see a wiring diagram detailing the various connections, particularly where it involves the electrical grid. A common way to show these wiring diagrams is by creating a simplified "single line diagram" such as the ones shown in Figure 13-19 a, b. You will have to include details such as wire sizes and type, fuse and circuit breaker ratings and any other details requested of the inspection authority.

The Generator

Generators supplied in North America are available with 120 V/240 V split phase or 120 V only output configurations. If your system is wired to provide 120 V only, having the generator supplier pre-wire it for 120 V will make installation simpler. Likewise, the same rules follow for 240 V systems.

The requirement of having all ground and neutral points bonded at one location (see Figure 13-7) will necessitate the removal of the ground-to-neutral connection inside the generator wiring chassis if the unit is part of a grid-interconnected system.

Phantom Load Supply

Earlier in the book we discussed the use of necessary phantom loads such as cell phone battery chargers, cordless phones, laptop computers, alarm systems and the like. The primary difficulty with phantom loads is a continual, slow draining of the battery bank without being aware of the effect. This is akin to coins falling out of your pockets.

The best way around phantom loads is not to have any. That being said, everyone has a few, so a simple method of working with them is to run these loads off of a very small, dedicated inverter such as the one shown in Figure 5-10 (and detailed schematically in Figure 13-19b). Provided the inverter is connected with a small number of loads, the system should accommodate your requirements. The outlet marked "special outlet" in Figure 13-19b is wired either to a central location where the phantom loads reside or a number of outlets may be wired in parallel throughout the house. As these outlets **must** be reserved for your special loads of very limited capacity, the outlet covers may be labeled, marked or otherwise identified to the satisfaction of the electrical authority.

Energy Meters

Earlier in the book we discussed how the electrical energy meter records the amount of electrons flowing into or out of the battery (current exported or imported). A device known as a shunt (see Figure 9-15) converts this current flow into a signal that can be measured by the meter. The shunt is typically mounted inside the breaker chassis box, ready for wiring in series with the battery bank negative terminal.

Figure 13-21 details the connection of the shunt. Refer to the specific energy meter wiring connection diagrams for the balance of system connection. Note that where the small gauge signal wire (typically #22 to #28 AWG) cable exits a hole in the breaker chassis, an anti-rubbing bushing must be added to prevent damage.

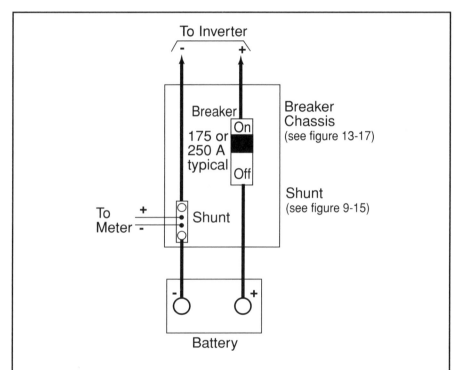

Figure 13-21. When ordering the main battery disconnect and circuit breaker, ensure that a 500 Amp shunt is included. This will allow connection to an energy meter either now, or at some future point with a minimum of hassle.

 *Sun-spiration*

"There are many myths attached to off-the-grid living. For instance, I have had people ask me if my lights work at night when you use solar power (why would you want lights that only worked during the day?) , if solar panels keep my house warm enough in winter (of course there is no connection here - we heat with wood) and if we have running water (we do).

I was talking to my daughter's teacher one day and told her about us being off-the-grid and using solar and wind to generate our electricity. She looked my daughters up and down and announced, "Your girls **look** clean for being off-the-grid."

I laughed and felt like assuring her that my daughters enjoy just as many hot, 30-minute long showers as any other teenager!"

                                                              Michelle Mather

Chapter 14
Living with
Renewable Energy

W ell, let's get ready to break out the champagne and "throw the switch". All the wiring is done, circuits are checked and re-checked, and the electrical inspector has given the OK. It's time to get things running. So where do you start? It can't really be that simple. After all, there is a lot of equipment installed.

Getting Started

The best place to start is with safety and bit of background on operating a system. Each source of renewable energy, whether it's the wood stove or PV panels, will have an operating guide supplied with it. All manufacturers will provide a list of safety tools and supplies that are necessary when working around their products. All of this equipment has been covered in the previous chapters, so go back and review the material as required. The most obvious rules are common sense and the ability to learn as you work with the various components.

Once all of the systems are checked out in accordance with the manufactures' requirements, we can start things moving. Obviously, the first thing to do is energize the system step-by-step. Note that some steps will not apply to grid-connected systems.

Step 1. Record the specific gravity for each battery cell prior to initial operation. This will provide a basis for determining the state of the batteries upon receipt and may be required if you have any warranty issues. Use a felt pen or marking plate to give each individual cell a number. If your system is 12 V, you will have 6 cells. Likewise 24 V and 48 V systems will have 12 and 24 cells per battery bank, respectively. If you have multiple sets of batteries wired in parallel, such those shown in Figure 13-14, you will have double the number of cells for each set.

Step 2. Energize the main circuit breaker connecting the batteries to the inverter. At this point, the inverter will be ready for startup. Follow the manufacturer's suggested startup procedure to get the unit online. You will need to set up the following basic parameters:
- battery charger rate
- battery bank capacity
- battery type
- over-discharge protection
- search mode sensitivity

More advanced inverters require setting many more options that can be done initially or on an as-needed basis.

Step 3. Set the search sensitivity if your inverter is equipped with this option. This setting adjusts the current threshold required to bring the inverter out of the low power "sleep" mode and into full power operation. The search mode works by sending a brief pulse of power into the power lines. When all circuits are turned off, the flow of electricity is stopped as we learned in *Chapter 2 – Energy 101*. At the instant the circuit is closed (when a light switch is turned on, for example), the current pulse travels through the load and back to the inverter, signaling it is time to activate. Sleep mode saves considerable electrical energy when the inverter is not required, for example , when everyone is sleeping.

Search mode can be tricky to adjust as the sensitivity differs depending upon which load is activated: an 8 W CF lamp uses far less power than a water pump. It is recommended that this adjustment be performed with a variety of electrical loads in the house, according to the inverter manual instructions.

Step 4. Program the energy meter according to the manufacturer's instructions. Typical data required will include the battery bank voltage and capacity (in amp-hours). With all renewable source supplies turned off, the inverter will be drawing power of varying amounts depending the electrical loads activated at that time. Try turning a light on or off to see the amount of current draw. Each load will require more or less current (in amps) causing the meter to register accordingly.

Step 5. Close the circuit breaker for the first renewable source supply. The input power will depend upon whether it is sunny and/or windy outside. The energy meter should now start to record the energy going into the battery. A common problem at startup is the current shunt wiring (Figure 13-21) being connected to the meter backwards. A quick test is to turn off all electrical loads and monitor the meter when energy is being produced. The meter should indicate a "battery charging" condition with the capacity moving slowly towards the full state. Repeat this step with each renewable source.

Step 6. Close the circuit breaker for the generator feed inputs. Start the generator either manually or by activating the auto-start function in accordance with the inverter instruction manual. Once the generator has started and stabilized for about 20 seconds, the inverter will "click over" to the charge battery mode.

When this occurs, the batteries should start to charge at the inverter's maximum rate (or as programmed). The energy meter will record the resulting power input and charging will progress.

While generator charging is in progress, verify that the house electrical power is active. All lights and appliances should be able to run as if they were running on the inverter. Note that the generator power may add a "flicker" condition to lights. This is completely normal operation.

Step 7: If the battery specific gravity is lower than 75% full upon receipt, as determined in Step 1, allow the generator to operate for one charging cycle as directed by the inverter manual. This will ensure that the battery electrolyte is up to 100% capacity as you begin to work with your new system.

Step 8. Take this time to read the energy meter manual to understand the relationship between the meter and batteries. Remember that the meter is only a "guesstimate" of actual battery state of charge.

Up and Running

The real beauty of a renewable energy system is its ease of operation once everything is commissioned and running. Ease of operation and proper attention to the system will prolong its life. Possibly the most neglected part of a system are the batteries. Those unfamiliar with renewable energy systems often do not understand how much energy they are using and the relationship between personal energy use and battery depth of discharge.

If there is any point that needs to be made clear from the beginning it is that you do not have a line of credit with your battery bank. There is only so much "cash" in the savings account. Use it up and that's it; it's lights out, literally. Keep draining the bank on a continuous basis and the bank will close permanently.

Read the battery manual to ensure you understand how your energy usage, energy meter and battery specific gravity work together. Once you fully understand the energy meter/specific gravity relationship, you only need to check the battery bank approximately once a month. But until then, use care and follow these steps to the letter:

- Battery specific gravity will read abnormally low immediately after a full day of charging. This is due to tiny gas bubbles suspended in the electrolyte causing the density of the fluid to decrease (gas bubbles are lighter than water or sulfuric acid). Wait at least one or two hours before taking the specific gravity reading.

- Likewise, specific gravity reading should not be taken when the batteries are under a full load. Wait until the water pump stops or microwave is not being used before taking readings.

- Battery electrolyte specific gravity readings that are outside room temperature range will need to be "adjusted" in accordance with the manufacturer's data.

- Use a quality hydrometer to take the readings. Look at the scale from straight on to ensure the reading is correct (Figure 9-3).

- Compare the specific gravity readings with the battery depth of discharge chart in Chapter 9, Table 9-1. If the battery bank indicates that it is 90% full (i.e. 10% depth of discharge) and the energy meter agree, great! If not, the meter may need some fine-tuning if the readings are too far out of whack. You may wish to look at a figure called the *charge efficiency factor* setting. This is a fancy term for the fudge factor to help calibrate the meter-to-battery setting. Decrease the efficiency factor if the meter thinks the batteries are more fully charged than they really are. Likewise, increase the efficiency factor if the meter reads lower than what the specific gravity reading indicates.

- As time goes by, the readings of even the best energy meter will fall out of step with the batteries. At this point, read the meter manual to determine how to reset the meter to agree with the battery capacity. This step is normally accomplished when the batteries are fully charged immediately after an equalization charge.

- Speaking of equalization, watch the specific gravity between cells. The readings should all be approximately the same. If the cells are starting to get out of balance by more than a reading of 0.010 (ten points on the scale), it's time to equalize.

- Equalization may be accomplished using a generator or better yet, using the PV panel or the wind turbine on sunny/windy days. Whatever method you choose, it is necessary to "program" the charge controller and/or inverter to start the equalization process. Many charge controllers have a single button to press, which increases the battery maximum charge voltage, thus enabling equalization mode. The inverter will often have a similar button or setting, which is activated when equalization is started from a generator.

- If you use Hydrocaps with your battery bank, remove them during equalization.

What Else?

There really isn't much else to think about, especially for on-grid systems. Equipment manufacturers will give you a list of yearly maintenance steps to follow. Some are absurdly simple such as spring-cleaning the PV cells to wash off any dirt on the glass. Wind turbines require an inspection, although even this can be pretty simple. Bergey WindPower suggests checking to see if the unit is turning once every year.... honestly, it's right there in the manual!

Of course there are some tricks you will learn over time that help make living with renewable energy easier:

- Like the old saying goes, "Make hay when it's sunny". If you have any big energy jobs to do, like 10 loads of laundry, better to do them when energy production is high. The same goes when using any high-energy appliances, such as:
  - vacuums
  - pool or hot tub pumps
  - air conditioning units
  - slow cookers

- If the generator is running during energy "dry spells", use that time to run the big power items listed above. Loading the generator to its highest capacity improves "gas mileage" and resulting generator efficiency.
- Watch the battery specific gravity or energy meter when you have a dry spell of two days or longer. This is where many newcomers to renewable energy get into trouble with their batteries. Recognize when the batteries are depleted enough to warrant running a generator. (Review Chapter 9 for more pointers.)
- During severe lightning storms, consider furling your wind turbine to limit stress damage during the storm. Consult the manufacturer's manual for additional details.
- Adjust the manual or single-axis tracker for seasonal operations as described in Chapter 6.
- Remember never to smash ice and snow off of the PV panels. Simply brush off the top layer of snow with a squeegee or brush. The sun will quickly take care of the rest.
- Rest easy knowing that you are not the only one living lightly on the planet. Support from dealers, like-minded neighbors, the "wwweb" and magazines such as Private Power and Home Power (www.homepower.com) abound. Renewable energy is here to stay.

Probably the most important issue of all is to enjoy your handiwork and marvel at the elegant simplicity of it all: the tracker follows the sun on its relentless journey across the sky appropriately aiming the PV panels and generating electricity. The wind whistles through the turbine blades making the turbine spin. A bunch of batteries and electronic stuff in the basement cranks out energy allowing Lorraine and me to enjoy our morning cappuccino in the hot tub, while watching the lake and enjoying the call of the loons. All of this made possible by our own off-grid, environmentally friendly power station.

Damn, life is good!

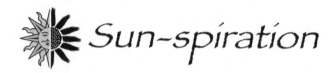 *Sun-spiration*

"My wife and I often reflect on the greatest accomplishments of our lives; from the joy of raising our children to the exhilaration of running our own business. Now, living off the power grid there is nothing as satisfying as going for a walk on a moonlit night (nature must cooperate for this, since the nearest streetlight is 14 kilometers away) and coming back to the sight of our little farmhouse, built in 1888, and surrounded by thousands of acres of forest, with no power lines running to it.

It glows with the warmth of the light that floods from our efficient, compact fluorescent light bulbs. Inside, our daughters are watching television, or doing homework on the computer, all without contributing to the smog that is becoming a reality throughout the country. Sometimes at school the next day they're a bit 'out of the loop'. Not because they live 6 kilometers from the nearest power pole, but because everyone else has the excuse of not getting their homework done, because there was a power outage the night before."

Cam Mather

Chapter 15
Conclusion

Renewable energy works. Yes, there is a lot of technology and things to understand before embarking on your quest for energy self-sufficiency, but the rewards are worth every effort. Lorraine tells me she feels more "connected" with her actions and interaction with the environment.

Once you starting harvesting your own energy and seeing what it takes to heat a house or run a light bulb, endless, mindless waste becomes intolerable. As we approach the end of 2003, hundreds of thousands of people enjoy a more sustainable lifestyle. This movement started with the hippies in the '60s, and after the psychedelic haze left that era, environmental concerns and technology started to enter the mainstream. It has now arrived.

Companies such as BP, Kyocera, Siemens and Xantrex are not in the business to help out aging hippies, they are here because renewable energy is business....big business. Governments, NGOs, industries as well as grassroots people like you and me are making the transition to a cleaner more sustainable future, for the sake of our children and the planet.

"To see a world in a grain of sand, and heaven in a wildflower; to hold infinity in the palm of your hand and eternity in an hour - is inspiration."

William Blake

Appendix 1
Cross Reference Chart of Various Fuels Energy Ratings

Fossil Fuels and Electricity

Heating Fuel	BTU per Unit
Heating Oil	142,000 BTU/gallon (38,700 kj/L)
Natural Gas	46,660 BTU/cubic-yard (37,700 kj/m3)
Propane	91,500 BTU/gallon (26,900 kj/L)
Electricity (resistance heating)	3413 BTU/kWh (3600 kj/kWh)
Coal (air dry average)	12,000 BTU/LB (27,900 kJ/kg)

Renewable Energy Heating Fuels

Heating Fuel	BTU per Unit
Shelled Corn	7000 BTU/lb (16,200 kJ/kg)
	14,000,000 BTU/ton (12,700 MJ/tonne)
Firewood by weight (all types) (note 1)	8000 BTU/lb (18,500 kJ/kg)
Hardwood Firewood by volume: (note1)	
Ash	25,800,000 BTU/cord (27,200 MJ/cord)
Beech	28,900,000 BTU/cord (30,500 MJ/cord)
Red Maple	22,300,000 BTU/cord (23,500 MJ/cord)
Red Oak	27,200,000 BTU/cord (28,700 MJ/cord)
Hybrid Poplar	18,500,000 BTU/cord (19,500 MJ/cord)
Mixed Hardwood (average)	27,000,000 BTU/cord (30,000 MJ/cord)
Mixed Softwoods (average)	17,500,000 BTU/cord (18,700 MJ/cord)
Wood Pellets	20,700,000 BTU/ton (19,800 MJ/tonne)
Biodiesel	128,000 BTU/gallon (35,500 kJ/L)

Note 1:
All firewood has the same heating or carbon content per pound or kilogram of mass. However, the density of softwoods are much lower owing to increased air and moisture content, resulting in lower BTU content per unit mass.

Appendix 2
Typical Power and Electrical Ratings of Appliances and Tools

Appliance Type	Power Rating (Watts)	Energy Usage per Hour, Day or Cycle
Large Appliances:		
Gas clothes dryer	600	500 Wh per dry cycle
Electric clothes dryer	6,000	5 kWh per dry cycle
Hi efficiency clothes washer	300	250 Wh per wash
10 yr. old vertical axis clothes washer	1,200	720 Wh per wash
10 yr. old refrigerator	720	5 kWh per day
New energy efficient refrigerator	150	1.2 kWh per day
10 yr. old chest deep freeze	400	3 kWh per day
New energy efficient chest deep freeze	140	0.9 kWh per day
Dish washer "normal cycle"	1,500	800 Wh per cycle
Dish washer "eco-dry cycle"	600	300 Wh per cycle
Portable vacuum cleaner	600	600 Wh per hour
Central vacuum cleaner	1,400	1.4 kWh per hour
Air conditioner 12,000 BTU (window)	1,200	1.2 kWh per hour
AC submersible well pump (1/2 hp)	1,150	200 Wh per cycle
DC submersible well pump	80	160 Wh per day
DC slow pump (includes booster pump)	80	160 Wh per day

Appliance Type	Power Rating (Watts)	Energy Usage per Hour, Day or Cycle
Small Appliances:		
Microwave oven (0.5 cubic foot)	900	0.9 kWh per hour
Microwave oven (1.5 cubic foot)	1,500	1.5 kWh per hour
Drip style coffee maker (brew cycle)	1,200	1.2 kWh per hour
Drip style coffee maker (warming cycle)	300	0.3 kWh per hour
Espresso/cappuccino maker	1,200	300 Wh per cycle
Food processor	300	50 Wh per cycle
Coffee grinder	100	10 Wh per cycle
Toaster	1,200	150 Wh per cycle
Blender	300	50 Wh per cycle
Hand mixer	100	10 Wh per cycle
Hair dryer	1,500	200 Wh per cycle
Curling iron	600	100 Wh per cycle
Electric tooth brush	2	50 Wh per day
Electric iron	1,000	1 kWh per hour

Appendix 2 Continued

Appliance Type	Power Rating (Watts)	Energy Usage per Hour, Day or Cycle
Electronics:		
Television –12 inch B&W	20	20 Wh per hour
Television –32 inch color	140	140 Wh per hour
Television –50 inch hi definition	160	160 Wh per hour
Satellite dish and receiver	25	25 Wh per hour
Stereo system	50	50 Wh per hour
Home theater system (watching movie)	400	1 kWh per movie
Cordless phone	3	72 Wh per day
Cell phone in charger base	3	72 Wh per day
VCR/DVD/CD component	25	25 Wh per hour
Clockradio (not including inverter waste)	5	120 Wh per day
Computer "Tower"	60	60 Wh per hour
Laptop computer	20	20 Wh per hour
15 inch monitor	100	100 Wh per hour
15 inch flat screen monitor	30	30 Wh per hour
Laser printer (standby mode average)	50	50 Wh per hour standby
Laser printer (print mode)	600	600 Wh per hour printing
Inkjet printer (all modes)	30	30 Wh per hour
Fax machine	5	120 Wh per day
PDA charging	3	72 Wh per day
Florescent desk lamp	10	10 Wh per hour

Appendix 3
Resource Guide

Energy Efficiency Councils and Societies:

American Council for an Energy Efficient Economy
Website: www.aceee.org
Phone: 202-429-8873
Publishes guides' comparing the energy efficiency of appliances

American Solar Energy Society
Website: www.ases.org
Phone: 303-443-3130
ASES is the United States chapter of the world Solar Energy Society. They promote the advancement of solar energy technologies.

The American Wind Energy Association
Website: www.awea.org
Phone: 202-383-2500
The AWEA is the trade association for developers and manufacturers of wind turbine and associated equipment and infrastructure.

California Energy Commission
Website: www.energy.ca.gov
Phone: 916-654-4058
The CEC is the strongest supporter of grid inter-connected renewable energy systems in North America. Their website explores what is happening in California in this regard.

Canadian Standards International
Website: www.csa-international.org
Phone: 416-747-4000
Develops standards for the Canadian marketplace. Tests and administers safety certification work in North America.

Canadian Wind Energy Association
Website: www.canwea.ca
Phone: 800-992-6932
The CAWEA vision is to have 10,000 MW of wind power systems installed in Canada by 2010. They promote all aspects of wind energy and related systems to the industry.

David Suzuki Foundation
Website: www.davidsuzuki.org
Phone: 614-732-4228
Dr. Suzuki is a lecturer and TV broadcaster promoting energy efficiency, global climate change and ocean sustainability. The website provides links and publications on all manner of environmental sciences.

Electro Federation of Canada
Website: www.micropower-connect.org
Phone: 905-602-8877
The Electro Federation is a consortium of electrical manufacturers working in many disciplines of electrical engineering and sales. The micropower-connect division is dealing with small (<50kW) distributed energy producers interconnecting to the grid in Canada.

Energy Star
Website: www.energystar.gov
Phone: 888-782-7937
Their website reviews energy efficient computers and electronics.

The Green Power Network
Website: www.eren.doe.gov/greenpower
This website describes the status of utility interconnection guidelines on a state-by-state basis. Also provides information on where to purchase green electricity when connected to the grid.

National Renewable Energies Laboratory (NREL)
Website: www.nrel.gov
Phone: 303-275-3000
The NREL is the national renewable energy research laboratory in the United States.

Natural Resources Canada
Website: www.nrcan.gc.ca
Phone: N/A
The Government of Canada hosts this website which includes the office of energy efficiency. Many resources are presented in this fact filled site.

Rocky Mountain Institute
Website: www.rmi.org
Phone: 970-927-3851
The Rocky Mountain Institute is a think tank regarding all energy efficiency issues. Their website contains a great deal of source information for books and applied research.

Underwriters Laboratories Inc.
Website: www.ul.com
Phone: 847-272-8800
Develops standards for the United States marketplace. Tests and administers safety certification work in North America.

Trade Publications and Magazines:

Home Power Magazine
Website: www.homepower.com
Phone: 800-707-6585
This magazine bills itself as "The hands-on journal of home-made power". Based in Oregon, the magazine deals primarily with a south and west coast flavor. Extensive details related to producing alternate energy.

Private Power Magazine
Website: www.privatepower.ca
Phone: 800-668-7788
Private Power is a new comer to the Canadian market place. Focuses on hybrid installations required in the North.

Alternative Power Magazine
Website: www.altpowermag.com
Phone: N/A
This web-based magazine offers general news-like stories related to all aspects of alterative power for home, automobiles and industry.

Manufacturers:
Photovoltaic Panels and Equipment
Photovoltaic panel manufacturers do not supply directly to end consumers, as they rely on their large distribution networks throughout the world. For information purposes, here are some of the major suppliers in this field. Contact the manufacturer or visit their website for distributors in your local area.

AstroPower
Website: www.astropower.com
Phone: 302-366-0400

BP Solar
Website: www.bpsolar.com
Phone: 410-981-0240

Evergreen Solar Inc.
Website: www.evergreensolar.com
Phone: 508-357-2221

Kyocera Solar Inc.
Website: www.kyocerasolar.com
Phone: 800-544-6466

Matrix Solar Technologies
Website: www.matrixsolar.com
Phone: 505-833-0100

RWE Schott Solar
Website: www.asepv.com
Phone: 800-977-0777

Schott Applied Power Corporation
Website: www.schottappliedpower.com
Phone: 888-457-6527

Sharp USA
Website: www.sharpusa.com
Phone: 800-BE-SHARP

Siemens Solar Inc.
Website: www.siemenssolar.com
Phone: 877-360-1789

Solardyne Corporation
Website: www.solardyne.com
Phone: 503-244-5815
Manufacturer of small solar modules for charging laptop computers, cell phones, etc.

Solarex
Website: www.solarex.com
Phone: 301-698-4200
Solarex was recently acquired by BP Solar

PV Module Mounts

Array Technologies Inc.
Website: www.wattsun.com
Phone: 505-881-7567
Manufacturer of the Wattsun active tracking system

Two Seas Metalworks
Website: www.2seas.com
Phone: 877-952-9523
Manufacturer of fixed PV racks. Also supply battery racks.

UniRac Inc.
Website: www.unirac.com
Phone: 505-242-6411

Zomeworks Corporation
Website: www.zomeworks.com
Phone: 800-279-6342
Manufacturer of fixed and passive tracking mount systems

Wind Turbines:

Atlantic Orient Corporation
Website: www.aocwind.net
Phone: 802-333-9400

Bergey Windpower Inc.
Website: www.bergey.com
Phone: 405-364-4214

Southwest Windpower
Website: www.windenergy.com
Phone: 520-779-9463

Aeromax Corporation
Website: www.aeromaxwindenergy.com
Phone: 888-407-9463

Bornay Windturbines
Website: www.bornay.com
Phone: +34-965-560-025
Manufacturer of wind turbines from Spain

Lake Michigan Wind and Sun
Website: www.windandsun.com
Phone: 920-743-0456
New and re-built turbines

Jack Rabbit Energy Systems
Website: www.jackrabbitmarine.com
Phone: 203-961-8133

Windstream Power Systems Inc.
Website: www.windstreampower.com
Phone: 802-658-0075

Wind Turbine Industries Corporation
Website: www.windturbine.net
Phone: 952-447-6064

True North Power Systems
Website: www.truenorthpower.com
Phone: 519-793-3290

Windturbine.ca
Website: www.windturbine.ca
Phone: 886-778-5069

Micro Hydro Turbine Systems
Energy Systems and Design
Website: www.microhydropower.com
Phone: 506-433-3151

Jack Rabbit Energy Systems
Website: www.jackrabbitmarine.com
Phone: 203-961-8133

Harris Hydroelectric
Website: www.harrishydro.com
Phone: 831-425-7652

Canyon Industries Inc.
Website: www.canyonindustriesinc.com
Phone: 360-592-5552

HydroScreen Co. LLC
Website: www.hydroscreen.com
Phone: 303-333-6071
Manufacturer of intake screen for micro-hydro systems

Battery Manufacturers

Dyno Battery Inc.
Website: www.dynobattery.com
Phone: 206-283-7450

HuP Solar-One Battery
Website: www.hupsolarone.com
Phone: 208-267-6409

IBE Battery
Website: www.ibe-inc.com
Phone: 818-767-7067

Rolls Battery Engineering (USA)
Surrette Battery Company (Canada)
Website: www.surrette.com
Phone: 800-681-9914

Trojan Battery Company
Website: www.trojanbattery.com
Phone: 800-423-6569

U.S. Battery Manufacturing Company
Website: www.usbattery.com
Phone: 800-695-0945

Hydrogen Recombining Caps

Hydrocap Catalyst Battery Caps
Website: N/A
Phone: 305-696-2504

D.C. Voltage Regulators

Morningstar Corporation
Website: www.morningstarcorp.com
Phone: 215-321-4457

RV Power Products
Website: www.rvpowerproducts.com
Phone: 800-493-7877

Steca Gmbh
Website: www.stecasolar.com
Phone: N/A

Xantrex Technology Inc.
Website: www.xantrex.com
Phone: 360-435-2220

Inverters

ExelTech Inc.
Website: www.exeltech.com
Phone: 800-886-4683

Out Back Power Systems
Website: www.outbackpower.com
Phone: 360-435-6030

SMA America Inc.
Website: www.sma-america.com
Phone: 530-273-4895

Xantrex Technology Inc.
Website: www.xantrex.com
Phone: 360-435-2220

Backup Power Gensets

Epower
Website: www.epowerchargerboosters.com
Phone: 423-253-6984

Generac Power Systems Inc.
Website: www.generac.com
Phone: N/A (sold through Home Depot)

Hardy Diesel & Equipment Inc.
Website: www.hardydiesel.com
Phone: 800-341-7027

Kohler Power Systems

Website: www.kohlerpowersystems.com
Phone: 800-544-2444

Energy Meters

Bogart Engineering

Website: www.borartengineering.com
Phone: 831-338-0616

Brand Electronics

Website: www.brandelectronics.com
Phone: 207-549-3401

Xantrex Technology Inc.

Website: www.xantrex.com
Phone: 360-435-2220

Miscellaneous

Bussmann

Website: www.bussmann.com
Phone: 314-527-3877
Fuses and electrical safety components

Delta Lightning Arrestors Inc.

Website: www.deltala.com
Phone: 915-267-1000

Digi-Key (Canada and USA)

Website: www.digikey.com
Phone: 800-DIGI-KEY
Supplier of many electrical wiring components

Electro Sonic Inc. (Canada)

Website: www.e-sonic.com
Phone: 800-56-SONIC
Supplier of many electrical wiring components

Real Goods

Website: www.realgoods.com
Phone: 800-919-2400
Suppliers specializing in renewable energy systems

Siemens Energy and Automation Inc.
Website: www.siemens.com
Phone: 404-751-2000
Fused disconnect switches and circuit breakers

Xantrex Technology Inc.
Website: www.xantrex.com
Phone: 360-435-2220
D.C. circuit breakers, battery cables, power centers, metering shunts, fuses

Appendix 4
Magnetic Declination Map for North America

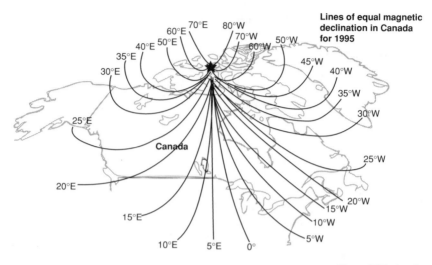

Lines of equal magnetic declination in Canada for 1995

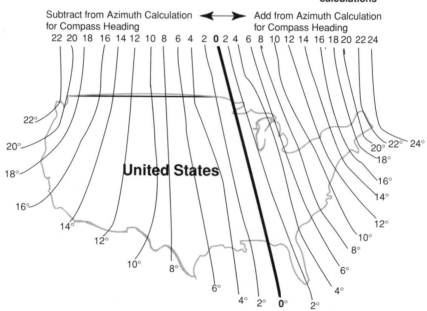

Map of US showing magnetic declatination for Azimuth compass calculations

Map indicates the correction heading required to locate true north (and solar south). If you live along the line running from Manitoba through to southern Texas, your heading will be due north when the compass is pointing approximately 8 degrees east.

Appendix 5
Winter Average Sun Hours per Day Map for North America

"North American Sun Hours per Day (Worst Month)"

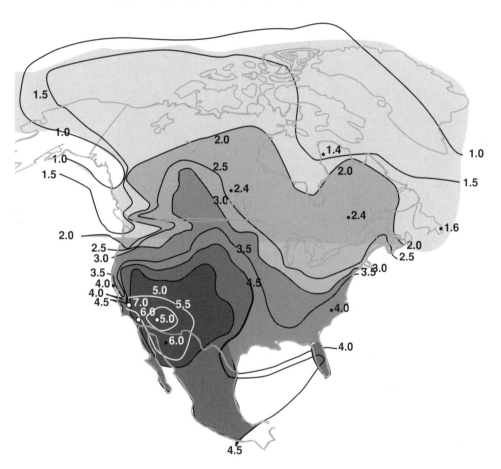

Map indicates the winter average of sun hours per day. Use this map when calculating average energy output of a PV system used all year round.

Appendix 6
Yearly Average Sun Hours per Day Map for North America

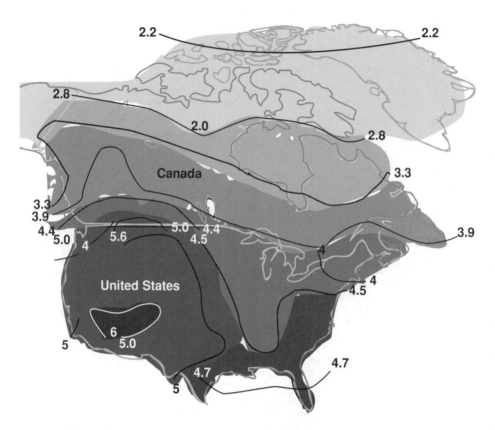

Map indicates the yearly average of sun hours per day. Use this map when calculating average energy output of a PV system used seasonally during the spring/summer months.

Appendix 7
Electrical Energy Consumption Worksheet

Appliance Type	Appliance Wattage (Volts x Amps)	X	Hours of Daily Use	=	Average Watt-hours Per Day

Total Watt-hours Per Day of all Appliances =

Start by checking your major electrical loads in the house. Anything that plugs into a regular wall socket will have a label that tells you the voltage (usually 120 or 240 for larger appliances) and current or wattage for that item. Although these labels can over-state energy usage, they are a good guide for calculating energy consumption. Write down the wattage data from the labels into the *Energy Consumption Worksheet*. (If the appliance label does not show wattage, multiply amps x volts to calculate watts).

The next step is to see if you can estimate how many hours you use the device per day. If you only use a device occasionally, try to estimate how long it is used per week and divide this time by 7. A quick fly through the calculator and voila', your "daily average energy usage per day" is computed.

Appendix 8
Average Annual Wind Speed Map for North America

Wind Class / Speed Chart
Class 1:   3.8 m / s (8.5 mph)
Class 2:   4.8 m / s (10.8 mph)
Class 3:   5.4 m / s (12.1 mph)
Class 4:   5.8 m / s (13.0 mph)
Class 5:   6.2 m / s (13.9 mph)
Class 6:   6.7 m / s (15.0 mph)
Class 7:   7.5 m / s (16.8 mph)

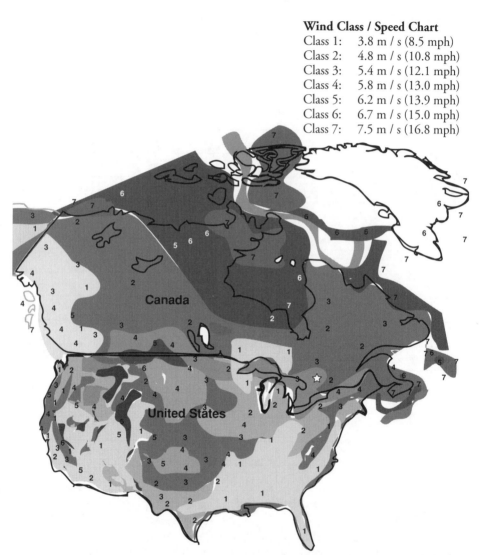

Map indicates the average annual wind speed in meters per second averaged over a ten-year period. Elevation of anemometer at 33 feet (10 m). Use this map with care as local wind speed levels will vary. Refer to resource guide for further details on wind mapping for your specific area.

Appendix 9a
Polyethylene SDR Pipe Friction Losses
Pressure Loss from Friction in Feet of Head per 100 Feet (30 m) of Pipe

(Table Courtesy of Energy Systems and Design)

Flow in US GPM	Pipe Diameter in Inches							
	0.5	0.75	1	1.25	1.5	2	2.5	3
1	1.13	0.28	0.09	0.02				
2	4.05	1.04	0.32	0.09	0.04			
2	8.60	2.19	0.67	0.19	0.09	0.02		
4	14.6	3.73	1.15	0.30	0.14	0.05		
5	22.1	5.61	1.75	0.46	0.21	0.07		
6	31.0	7.89	2.44	0.65	0.30	0.09	0.05	
7	41.2	10.5	3.24	0.85	0.42	0.12	0.06	
8	53.1	13.4	4.14	1.08	0.51	0.16	0.07	
9		16.7	5.15	1.36	0.65	0.18	0.08	
10		20.3	6.28	1.66	0.78	0.23	0.09	0.02
12		28.5	8.79	2.32	1.11	0.32	0.14	0.05
14		37.9	11.7	3.10	1.45	0.44	0.18	0.07
16			15.0	3.93	1.87	0.55	0.23	0.08
18			18.6	4.90	2.32	0.69	0.30	0.09
20			22.6	5.96	2.81	0.83	0.35	0.12
22			27.0	7.11	3.36	1.00	0.42	0.14
24			31.7	8.35	3.96	1.17	0.49	0.16
26			36.8	9.68	4.58	1.36	0.58	0.21
28				11.1	5.25	1.56	0.67	0.23
30				12.6	5.96	1.77	0.74	0.25
35				16.8	7.94	2.35	1.00	0.35
40				21.5	10.2	3.02	1.27	0.44
45				26.8	12.7	3.75	1.59	0.55
50				32.5	15.4	4.55	1.91	0.67
55					18.3	5.43	1.96	0.81
60					21.5	6.40	2.70	0.94
65					23.8	7.41	3.13	1.08
70					28.7	8.49	3.59	1.24
75					32.6	9.67	4.07	1.4
80						10.9	4.58	1.59
85						12.2	5.13	1.77
90						13.5	5.71	1.98
95						15.0	6.31	2.19
100						16.5	6.92	2.42
150						34.5	14.7	5.11
200							25.0	8.70
300								18.4

Appendix 9b
PVC Pressure Class 160 PSI Pipe Friction Losses
Pressure Loss from Friction in Feet of Head per 100 Feet (30 m) of Pipe
(Table Courtesy of Energy Systems and Design)

Flow in US GPM	Pipe Diameter in Inches							
	1	1.25	1.5	2	2.5	3	4	5
1	0.05	0.02						
2	0.14	0.05	0.02					
2	0.32	0.09	0.04					
4	0.53	0.16	0.09	0.02				
5	0.80	0.25	0.12	0.04				
6	1.13	0.35	0.18	0.07	0.02			
7	1.52	0.46	0.23	0.08	0.02			
8	1.93	0.58	0.30	0.10	0.04			
9	2.42	0.71	0.37	0.12	0.05			
10	2.92	0.87	0.46	0.16	0.07	0.02		
11	3.50	1.04	0.53	0.18	0.07	0.02		
12	4.09	1.22	0.64	0.20	0.09	0.02		
14	5.45	1.63	0.85	0.28	0.12	0.04		
16	7.00	2.09	1.08	0.37	0.14	0.04		
18	8.69	2.60	1.33	0.46	0.18	0.07		
20	10.6	3.15	1.63	0.55	0.21	0.09	0.02	
22	12.6	3.77	1.96	0.67	0.25	0.10	0.02	
24	14.8	4.42	2.32	0.78	0.30	0.12	0.04	
26	17.2	5.13	2.65	0.90	0.35	0.14	0.05	
28	19.7	5.89	3.04	1.04	0.41	0.16	0.05	
30	22.4	6.70	3.45	1.17	0.43	0.18	0.05	
35		8.90	4.64	1.56	0.62	0.23	0.07	
40		11.4	5.89	1.98	0.78	0.30	0.09	0.02
45		14.2	7.34	2.48	0.97	0.37	0.12	0.04
50		17.2	8.92	3.01	1.20	0.46	0.14	0.04
55		20.5	10.6	3.59	1.43	0.55	0.16	0.05
60		24.1	12.5	4.21	1.66	0.64	0.18	0.07
70			16.6	5.61	2.21	0.85	0.25	0.09
80			21.3	7.18	2.83	1.08	0.32	0.12
90				8.92	3.52	1.36	0.39	0.14
100				10.9	4.28	1.66	0.48	0.18
150				23.2	9.06	3.5	1.04	0.37
200					15.5	5.96	1.75	0.62
250					23.4	9.05	2.65	0.94
300						12.6	3.73	1.34
350						16.8	4.95	1.78
400						21.5	6.33	2.25
450							7.87	

Appendix 10
Voltage, Current and Distance (in feet) Charts
for Wiring (AWG gauge)
(1% Voltage drop shown – Increase distance proportionally if greater losses allowed)

12 Volt Circuit

Amps	WIRE GAUGE									
	10	8	6	4	2	1	1/0	2/0	3/0	4/0
1	48	74	118	187	299	375	472	594	753	948
5	10	15	24	37	60	75	94	119	151	190
10	5	7	12	19	30	38	47	59	75	94
20	2	4	6	9	15	19	24	30	38	48
40	1	2	3	5	7	9	12	15	19	24

24 Volt Circuit

Amps	WIRE GAUGE									
	10	8	6	4	2	1	1/0	2/0	3/0	4/0
10	10	15	24	37	60	75	94	119	151	190
20	5	7	12	19	30	38	47	59	75	94
30	3	5	8	12	20	25	31	40	50	63
40	2	4	6	9	15	19	24	30	38	48
50	2	2	4	6	10	13	16	20	25	31
100	1	1	2	4	6	8	9	12	15	19
125	1	1	2	3	5	6	8	10	13	16
150	1	1	2	2	4	5	6	8	10	13

48 Volt Circuit

Amps	WIRE GAUGE									
	10	8	6	4	2	1	1/0	2/0	3/0	4/0
10	20	30	48	74	120	150	188	238	302	380
20	10	15	24	38	60	76	94	118	150	188
30	6	10	16	24	40	50	62	80	100	126
40	4	8	12	18	30	38	48	60	76	96
50	4	4	8	12	20	26	32	40	50	62
100	2	2	4	8	12	16	18	24	30	48
125	2	2	4	6	10	12	16	20	26	32
150	2	2	4	4	8	10	12	16	20	26

Appendix 11
Wire Sizes Versus Current Carrying Capacity
(Size shown is for copper conductor only)

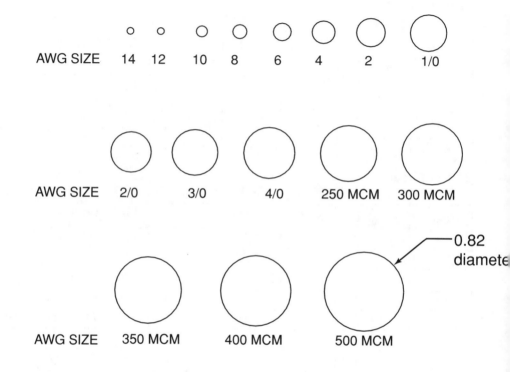

AWG	Maximum Current
14	15
12	20
10	30
8	55
6	75
4	95
2	130
0	170
2/1	195
3/0	225
4/0	260

Appendix 12
Communications Off-The-Grid

Although communications off the grid is not actually a renewable energy issue, it is an important consideration in planning a rural home. Cam Mather runs his electronic publishing company (the company that produced this *Handbook*) many miles from the nearest hydro and phone lines. Uploading graphics, emails and simply talking to clients is not a luxury, but a necessity to him. As someone concerned with long distance communications, Cam has provided his insight into this technological forest of options:

While living free of the electricity grid can be extremely gratifying, communicating with the outside world can be a challenge. This is because the power poles that bring electricity to homes, usually bring other services, including cable television and telephone.

If the utility poles don't reach your house, and you want the convenience of a telephone, you have several options.

Cell Phones:

With the spread of cellular service (Analogue 900 Mhz or Digital 1.9 Mhz) across North America, this should be the first option you look at. Does a cellular provider cover your area, and how good is the service? Remember most cellular providers are trying to minimize their investment in infrastructure, so they will tend to concentrate service in urban areas, and along major highway corridors. You may get great service as you pull off the highway on your way home, but service may be spotty at best when you arrive home.

When cellular service was first introduced (analogue), most people's phones were about 3 watts. These typically would be installed in a car with a fixed antenna on the back windshield, or come in the form of a bag phone with a small portable antenna.

Then the small, hand held, digital phones which produce only a half a watt of power became popular. Since most of the users were in urban areas, or wanted to use them while on highways, it became more difficult to find the higher wattage phones. The greater the power output of your phone, the more likely you'll be able to link up with a cellular antenna tower in your area. Your first task will be to find which cellular provider offers the best service in your area and then try and track down the highest wattage phone you can. Inquire about analogue 800 Megahertz transceivers. Lower frequency units have greater range than higher frequency models.

Cellular service works by having a series of service coverage areas or "cells" that are provided by an antenna tower. As you travel on a highway for instance, you may notice your phone service becomes weak or more static is heard as you are switched between local and distant cell towers.

In a rural location, you may find many or all of your calls are long distance. This is because you may be quite a distance from an antenna tower and the serving areas do not match the wireline services. If you are using an omnidirectional antenna rather than a fixed yagi, you may also notice on your phone bill that the calls seem to be charged as if you are driving around the area, even though all your calls have been made from your home. This is because when you placed the call, the system looked for the nearest antenna tower that had the capacity to handle your call. Theoretically each time you call, you may get bumped from one antenna tower to another, and thus it will look on your bill as if the calls have originated from a different location, when in fact it is just reflecting the actual antenna tower that your call was connected to.

If you find this to be an inconvenience you may want to try and work with a local provider to improve the quality of service. This could be in the form of you asking them to boast the signal in your area (unlikely but worth a try) or installing a fixed antenna (Yagi - directional antenna) on your cell phone that is directed at the nearest cellular antenna. This will offer you the greatest potential for cellular service.

Point-to-point phone line service/Phone Line Extender

(Lower frequency in the VHF 150 Mhz - UHF 400 Mhz to 1 Gigahertz range)
If you live in an area not covered by cellular service, but you have a phone line within 20 miles (32 kilometers) of your home, then you may want to look at a point-to-point phone line system, also called a phone line extender. This system will involve positioning a unit at the end of the phone line that converts the phone signal to a radio frequency. You will have a similar device at your home that will convert the radio frequency back to a phone signal, so your phone, fax and computer modem doesn't even notice the difference. You will in fact be using two radio frequencies (duplex), one to send information and one to receive.

This type of system will generally require you to be licensed on those frequencies since radio frequencies are considered public property. In Canada, Industry Canada licenses these radio frequencies and in the United States it is the FCC. You will be required, like a radio station, to pay a yearly fee to license these frequencies.

These systems have advantages and disadvantages.

Advantages

As long as you are close enough in terms of your phone exchange, chances are good a phone call to the nearest town may be a local call, saving access fees, especially if cellular service in your area results in calls being billed at long distance rates. The local calling area is determined by where your land line is connected to.

These systems mimic a real phone line, so you'll be able to use your fax machine and your computer's modem to log onto the Internet, etc. This can be expensive if you rely exclusively on cellular service.

Disadvantages

The purchase price of these systems can be significant. These systems typically cost $3,000 to $7,000 or more to install. If you are running a business from your home and can amortize this over time, it may not be a concern. On the other hand if you are hoping to get "back to basics" in your new off-grid lifestyle, a system like this will remind you that technology that may allow this move has a cost associated with it.

The "bandwidth" on some systems will be limited, which means that your Internet service may be slow. It can only be a good as the telephone line. If you just use it as a phone line, or to check your emails, it will be fine. If you want to download large files, this system may be prohibitively slow.

Since you are using radio frequencies, it will be subject to the whims of the weather. Cloudy or wet weather may impede its performance, just like a radio station that comes in clearly on sunny days, may have static on cloudy days.

You will need a location to install the transmission box and antenna at the end of the phone line. Your local phone company may allow you to install it on one of their poles (not likely), but they may not be able to in all instances. If the end-of-the-line telephone poles are on private property, you will require permission from the landowner.

Installation

These systems will require an antenna, one that should clear the tree line and be able to "see" the distant antenna. The higher the frequency, the more important it becomes to have a line of site between the two points, whereas at VHF or UHF you may not need a good line of site if

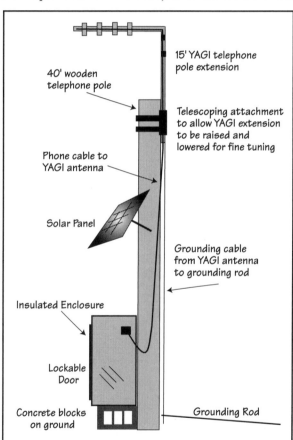

Figure A12-1. The remote transceiver end of Cam Mather's phone system is shown in this simplified schematic view. Note the YAGI antenna at the top of the pole, which is aimed at the receiver unit 20 miles (32 kilometers) away.

you are relatively close.

A common antenna type is a "YAGI" directional antenna, which will allow you to aim the antenna at your home directly from the remote one. Your antenna tower can be as basic as a wooden telephone pole or television antenna with the YAGI adapter attached at the top; as long as it clears the major obstacles between it and the other antenna.

At your home, you'll be able to power the transceiver as you would any appliance, be it AC or DC. (This is an excellent example of a necessary phantom load.)

At the remote location you will have to deal with how to power the unit. If it is located in a neighbor's home, power should not be a problem. But if you decide to locate in a remote utility box, you'll need to find a source of power for it. Chances are you'll have power there from your local electrical utility, because where there are phone lines there will be electricity on the line as well. You'll have to contact the electric utility and arrange to pay the expense for this service, which could be similar to having electricity brought into a home. You'll also be subject to power outages and service interruptions on your phone line.

Insulated Enclosure for transmit unit and solar power/batteries for Optaphone

Inside dimensions are 28" wide, 20" tall, and 9" deep
Back panel should allow mounting of voltage regulator
Has a locking handle with door to prevent vandalism
Must be mouse/animal proof

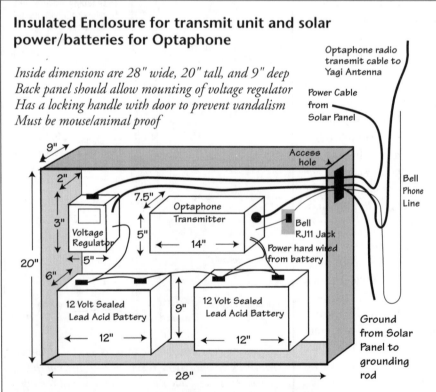

Figure A12-2. Mechanical view of the insulated enclosure for transceiver unit and solar power/battery pack.

Figure A12-3 (above). Cam Mather with the transceiver unit and battery pack opened.

Figure A12-4 (right). The transceiver unit is mounted in a lockable chassis at the end of the phone line. Note the small PV panel at the top of the pole used to charge the radio battery pack.

An alternative is to make your own power, just as you are doing at your home. The communication company that sells you the phone system should be able to help you with this. In the installation outlined in the illustrations, power is provided by one 75-Watt solar panel which charges two gel-cell sealed lead acid batteries. This box has no source of heat, so during the winter months although the box has some rigid Styrofoam insulation in it, it will get quite cold. This installation was based on the system being able to store enough DC power in the batteries to run the phone system for 4 hours per day for two weeks, with no sun, during the Canadian winter months (when the batteries will be cold and not perform as well). In the 4 years it has run, sometimes well in excess of four hours per day, there has always been enough power.

In the utility box you can see the series type solar charger similar to one used in a home, as well as the service box where the phone company brings in their line. The YAGI antenna is grounded to two 8-foot grounding rods. Notice how all 3 lines that come into the box, the phone line, the power line from the solar panel and the phone system antenna, all form a "U" or a drip line before they enter the box. This allows water running down the lines during a rainstorm to "drip" off rather than run into the electrical equipment.

Keeping water out of this utility box is absolutely essential. During a major rain storm, some rain was able to enter via the air vent on the top. We have since corrected this, and although it wasn't much, it was enough to drip onto the power cable connected to the Optaphone. Eventually the wires inside the plastic cover rusted, and shorted out. The problem was that the short was not visible. We were many days without a phone before we diagnosed what seemed to be such a minor problem. Water and electricity don't mix!

Point-to-point phone line service/Phone Line Extender
(Higher frequency in the 1 Gigahertz to 2.5 Gigahertz range)
Similar to the above installation, the main difference with these newer systems is that they work at a higher frequency and smaller objects will interfere with the signal. For instance with a system operating at 400 Mhz, if your antenna isn't high enough, some of the signal is still able to pass through the trees and operate properly.

In the 2.5 Ghz range however, obstacles as small as large rain drops can interfere with the signal, so it will be imperative you have complete "line of sight" between the two antennas. That means if you're at the top of one of the antennas you should be able to see the top of the other antenna with nothing impeding the line of site.

The advantage of these newer systems is that in this frequency range you are not licensed for frequency usage. This will save a yearly license fee, which is approximately U.S. $140 per year in Canada. However the cost of the additional tower height at both ends will probably off set this savings and you are prone to more interference. These units also provide greater "band width" which will allow faster Internet service. Remember they will only be able to provide internet speeds as fast as you would get at the end of the phone line, and rural service is often much slower than an urban area. The further you are from a phone substation, the slower your Internet access will be.

Companies that provide these products are:

Carlson Wireless Technologies
http://www.carlsonwireless.com/
1-707-923-3000

Exicom International
http://www.exicom.co.nz
64 4 237 0169

In Canada you can contact
Cartel Communication Systems Inc.
www.cartelsys.com
800-663-0070

Fixed-Point Broadband CDMA (Digital) Wireless Service From Your Phone Company

Similar to the point-to-point service, your local phone company may actually offer a similar service but in this case the system can service multiple customers. The advantage to the phone company is that it will save the cost of stringing copper cable along utility poles to customers. As the cost of this technology comes down, and the cost of installing utility poles goes up, you may find your phone company receptive to provide such a service.

In a rural area where there are no existing utilities poles, the phone company may reject your request for phone and suggest it is too expensive to provide. First off, you should admit they have a point, since you choose to live in a more rural area and receive the benefits it brings. If you want cheap, high-speed Internet service and phone service, you may have to move to a big city.

High Speed Internet Service from a Satellite Provider

Many rural homeowners are used to receiving outstanding television reception and channel selection through their satellite dish and receiver. Some satellite television companies offer Internet "down" service from their satellites. However, the user must still use their phone line to "upload" data to the Internet Service Provider (ISP). (Note that as of this writing several satellite providers are installing bi-directional Internet service).

For example, if you are on the Internet and wanted to download a form, the file would be broadcast to your computer from the satellite via a special modem installed in your computer. If you filled out the form and wanted to send it back to their website, you would use your phone line "dial up" connection.

The alternative is to use a specialized high-speed, satellite bi-directional Internet service. You will need a new satellite dish and support equipment. There will be some expense in purchasing this equipment initially, and a higher monthly rate, but you'll be receiving high speed internet without tying up your phone line, and chances are it will be an excellent service for you.

As the technology improves you may ultimately be able to get "Voice over IP". This is the term used to describe using the Internet as a phone system. If your Internet service is fast enough, you may actually be able to use it to transmit your voice, and speak to another Internet user who has the same capability.

If your satellite service provides unlimited Internet access, you'll also have un-limited long distance as well. This technology is new and there are many challenges in moving voice over the Internet, especially if it's originating from a satellite-based system. Ultimately, you may be able to use this as an alternative to some of the other phone options.

In the United States and Canada there are several companies that offer this service:

Linc-Sat
www.linCsat.com
1-877-546-2728

C-COM Satellite Service
www.c-comsat.com
1-877-463-8886

DirecPC
http://www.direcPC.com

Satellite Phone Connection

If your dream home is in the wilds of Alaska one further communication system to consider is the satellite phone. You may already be aware of these units installed in the seat backs of many airlines.

These remarkable phone units are capable of operating anywhere in the world and provide Internet connection as well. Another amazing feature of these systems is their ability to drain your bank account. One model of phone was recently priced at $700. The airtime fees ranged from $1.20 per minute for a 30-minute block to $300 per month for the 1250-minute "economy" package.

I am sure Bill Gates has a couple of these babies lying around the Yacht!

Conclusion

If you are looking to move away from utility lines, make sure you have a phone strategy and ensure it's working before you take the plunge. More and more people work out of their homes, or are dependent on phone and Internet service for many of their daily activities, from banking, to paying bills. While the dream to "drop out" to Walden Pond remains an admirable one, most of us still require some inter-action with the industrial economy and for most of us; the phone system is an integral part of that link.

Handbook Index